An introduction to
the ionosphere and magnetosphere

An introduction to the ionosphere and magnetosphere

J. A. RATCLIFFE

Cambridge
At the University Press
1972

Published by the Syndics of the Cambridge University Press
Bentley House, 200 Euston Road, London NW1 2DB
American Branch: 32 East 57th Street, New York, N.Y.10022

Library of Congress Catalogue Card Number: 74-171680

ISBN: 0 521 08341 9

Printed in Great Britain
at the University Printing House, Cambridge
(Brooke Crutchley, University Printer)

Contents

[v]

Preface

As a preparation for the writing of this book I have been privileged, for many years, to have discussions and arguments with many colleagues, first at Cambridge and then at Slough. I want to thank them all and especially Dr K. G. Budden, Dr P. C. Clemmow, Dr J. P. Dougherty, Dr E. C. Dunford, Dr H. Rishbeth, Dr L. Thomas, and Dr D. M. Willis, who have always been particularly ready to discuss the intricacies of the subject with me. I am also deeply indebted to Professor R. L. F. Boyd who has given me much help, particularly with chapter 10. Those who have helped me may not fully agree with my unorthodox treatment of some of the topics, and, of course, if there are errors they are mine. The design of the book has altered considerably since I began writing and there have been many versions; to Mrs Joan Reynolds, who has typed them all with exemplary speed and accuracy, my very best thanks are due.

J. A. RATCLIFFE

Cambridge
April 1971

Introduction

The first part of this book shows how the structure and behaviour of the ionosphere and magnetosphere are determined by the nature of the atmospheric gases and of the solar radiations that fall upon them. The second part outlines the physical principles that govern some of the processes occurring in the ionosphere and some of the experimental methods of investigating it by radio waves or by apparatus in space vehicles.

The book is not intended to be a comprehensive textbook or a detailed research monograph: but rather a guide that will show the reader something of the physical processes at work in the highest parts of the earth's atmosphere. The processes are numerous and their interactions complicated: the way in which they combine to produce the observed behaviour is often in doubt. Here no attempt is made to describe or to explain that behaviour in detail; the emphasis is on the nature of the physical processes themselves, not on their relative importance at any particular time or place.

Although most of the ionospheric processes mentioned in part 1 are fundamentally simple, a full discussion of the detailed complications that can arise is necessarily involved; it is usually presented as a somewhat forbidding piece of mathematics. One who wishes primarily to understand the principles involved, but not all the details, does not usually need to follow all the mathematical complications. In part 2, therefore, an attempt is made to present, in a simple way, the physical principles that govern some of these processes. The presentation, although believed to be sound, is not always that of the 'accepted' theory. A worker who intends to study a part of the subject in detail will need to acquaint himself with the standard approach, but it is hoped that he will understand it the better if he has first thought about the matter in the way described here. Chapters 9 and 10 (in part 2) outline the physical principles underlying the action of the most important of the experimental methods used in ionospheric investigations. Again the emphasis is on principles, not on the details of apparatus.

In discussing the whole of the ionized upper atmosphere there is a problem of nomenclature. Many workers consider that the lower part should be called the ionosphere and the upper part the magnetosphere. If these names were used it would be necessary in many places to refer awkwardly to 'the ionosphere and magnetosphere' when the whole was intended; moreover there is no agreement about the height of the boundary that separates the two. Because of this difficulty the nomenclature recommended by the Institute of Radio Engineers in their *Standards on Wave Propagation: Definition of Terms* (reference 158) is used, and the *ionosphere* is defined as 'that part of the earth's atmosphere in which free electrons exist in sufficient numbers to have an important effect on the travel of radio waves'. The word ionosphere is thus used to include the whole of the earth's environment between a height of about 60 km and the magnetopause. The word *magnetosphere* is then used, somewhat loosely, to mean 'the region in which the magnetic field of the earth has a dominant control over the motions of the charged particles'. This definition is similar to that of Gold [92] except that he considered the magnetosphere to be separate from the ionosphere, whereas here the magnetosphere is taken to be part of the ionosphere.

To help in detailed study of subjects that are dealt with only in outline, a bibliography is provided and reference is made to it by numbers in brackets. It is divided into two sections, the first containing books and general articles that extend some of the present discussions and that contain more detailed references, the second containing key papers that give detailed accounts of specialized subjects.

The angular (radian) frequency of an oscillation is referred to many times, but, to be concise, it is called simply the frequency except where it might be confused with the frequency measured in complete cycles. In accord with standard practice in the United Kingdom SI units are used throughout. Factors that convert them to other units are listed in appendix C, which also contains a list of the more important symbols used.

Part 1

The formation and nature of the ionosphere and magnetosphere†

1 The neutral atmosphere modified by the sun's radiation‡

1.1 Introduction

This book is concerned with the free electrons that are produced at great heights by the ionizing action of the sun's radiation on the earth's atmosphere. The part of the atmosphere above about 60 km, where free electrons exist in numbers sufficient to influence the travel of radio waves, is called the *ionosphere*. The earth's magnetic field affects the electrons' motion at all heights but with increasing importance at greater heights; that part of the ionosphere where it exerts a predominant influence is called the *magnetosphere*. Although there is no clear height at which the magnetosphere can be said to start, it is suggested that its base might be taken to be at the plasma-pause (see p. 75). To show how the electrons in the ionosphere are produced by the solar radiation, it is first necessary to discuss the nature of the neutral atmosphere.

1.2 The gases of the atmosphere

Near the ground the atmosphere contains 78 per cent of molecular nitrogen and 21 per cent molecular oxygen, by volume, together with several very secondary constituents, of which helium (4×10^{-4} per cent) is of interest for the purpose of this book. Up to a height of approximately 100 km the gases are mixed by turbulence and their relative proportions remain constant. At greater heights, where there is no turbulence, each is distributed as though it alone were present; it is

† [4, 20, 27, 28, 32, 34].   ‡ [7, 11, 125, 173].

[3]

said to be in diffusive equilibrium. The height at which turbulent mixing is replaced by diffusive equilibrium is called the *turbopause*.

1.2.1 Diffusive equilibrium

Each elementary volume of the gas rests in equilibrium under the downwards force of gravity and the difference between the gas pressures (p) on its lower and upper faces. If there are n molecules of mass m in each unit volume, and if the temperature T is uniform then, with k = Boltzman's constant and g = acceleration of gravity, $p = nkT$ and

$$nmg = -\frac{dp}{dh} = -kT\frac{dn}{dh} \tag{1.1}$$

so that

$$\frac{1}{n}\frac{dn}{dh} = -\frac{mg}{kT} \tag{1.2}$$

If g is assumed to be independent of height, (1.2) leads to

$$n = n_0 \exp(-h/H) \tag{1.3}$$

where

$$H = kT/mg \tag{1.4}$$

is called the *scale height* and n_0 is the concentration at some reference level, conveniently the turbopause.

Equations (1.2) and (1.4) show that the scale height is given by

$$\frac{1}{H} = -\frac{1}{n}\frac{dn}{dh}; \tag{1.5}$$

it provides a convenient description of the height-gradient of the concentration. Some confusion may arise in the following way through the use of the term 'scale height'. In the previous discussion it was defined by $H = kT/mg$ and it appeared as the constant H in the expression $\exp(-h/H)$ for the height-variation of the concentration. But it is sometimes convenient to discuss an exponential height-variation represented by $\exp(-h/\delta)$ where δ is not equal to H. Examples are: the equilibrium height-distribution of ions that are being produced and destroyed at rates that depend on height in different ways (p. 45); or the height-distribution of a gas that is not in equilibrium (p. 107). In describing situations like these the name *distribution height* will be given to the quantity δ: the name scale height will be reserved for the quantity defined by $H = kT/mg$.

Below the turbopause, where the gases are completely mixed, all have the same height-variation: it is given by (1.3) and (1.4) if m is the average mass of the molecules (or atoms). Near the ground the distribution height (of all constituents whatever their scale heights) is about 8 km. Above the turbopause, in the region of diffusive equilibrium, each gas is distributed with its own distribution height, usually equal to its scale height. The lightest gas, hydrogen, thus predominates at the greatest heights. This gas cannot be detected at ground level; it is probably formed at heights around 100 km by the action of solar radiation on water vapour.

The simple expression (1.3) is based on the supposition that g and the temperature (T) are independent of height. Although, at heights comparable with the radius (R) of the earth, it can no longer be supposed that g is constant, equations (1.3) and (1.4) are still correct if height (h) is measured in terms of a 'geopotential height' h^* defined by

$$h^* = Rh/(R+h)$$

Because the sun's radiation heats the atmosphere the temperature varies with height and with time. Fig. 1.1 shows extreme, and average, forms of this height-variation [7]. When the temperature is height-dependent (1.1) takes the form

$$nmg = -\frac{dp}{dh} = -k\left\{T\frac{dn}{dh} + n\frac{dT}{dh}\right\} \tag{1.6}$$

or, if $kT/mg = H$ represents the height-dependent scale height,

$$n = -H\frac{dn}{dh} - n\frac{dH}{dh} \tag{1.7}$$

In the simple case where there is a uniform gradient of temperature, and with it of scale height, so that

$$H = H_0 + \beta h, \tag{1.8}$$

integration of (1.7) leads to

$$\frac{n}{n_0} = \left(\frac{H}{H_0}\right)^{-(1+1/\beta)} \tag{1.9}$$

$$= \left(1 + \frac{\beta h}{H_0}\right)^{-(1+1/\beta)} \tag{1.10}$$

Whatever be the height-distribution of a particular gas its partial pressure p_0 at any height h_0 is equal to the weight of that gas contained in a volume of unit cross-section above that height, so that, if there are N_0 particles of mass m in the column, $p_0 = N_0 mg$. But $p_0 = n_0 kT_0$ where n_0 and T_0 are the concentration and the temperature of the gas particles at the level h_0; hence

$$N_0 = n_0 kT_0/mg = n_0 H_0 \qquad (1.11)$$

where H_0 is the scale height of the gas at the level h_0. A knowledge of the particle concentration and the scale height of any particular gas

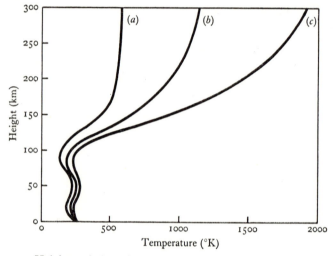

Fig. 1.1. Height variation of atmospheric temperature when heating by solar radiation is (*a*) weak – night, (*b*) average – day, and (*c*) strong – day at maximum of the solar cycle (after reference 7).

at any one level thus provides a knowledge of the total number of particles of that gas above that level, independent of how they are distributed in height.

1.2.2   The exosphere. Escape of hydrogen and helium
[71, 98, 111, 126, 143]

The ordinary gas laws can be applied to atmospheric gases only so long as the molecules make enough collisions to establish statistical equilibrium with their neighbours. If a molecule were to move vertically without collisions, and with velocity appropriate to temperature T, it would rise to a height equal to $3kT/2mg = 3H/2$. If this distance, of order equal to its scale height, were less than the mean free path (l) a molecule would, on the average, make no collisions and it could be considered to move freely. The part of the atmosphere where free movement of this kind occurs is called the *exosphere*; in it $l > H$. The mean free path (l) is given, approximately, in terms of the collision cross-section (σ) and the concentration (n) by the expression $l = (\sigma n)^{-1}$; hence at the base of the exosphere, where $l = H$, the relation $Hn = \sigma^{-1}$ applies. But from (1.11) $Hn = N$, the total number of particles in a superincumbent column of unit cross-section, so that the base of the exosphere is at a height where $N\sigma = 1$.

Most particles entering the exosphere from below travel in gravity-controlled orbits, without making collisions, until they either escape from the earth or re-enter the atmosphere below. If a particle is to escape from the earth's gravitational field it must have a kinetic energy greater than the gravitational potential energy at the radial distance r of the exospheric boundary so that, if m is its mass and v its velocity, $\frac{1}{2}mv^2 > mg_r r$ where g_r is the acceleration of gravity at the boundary. v is thus given by

$$v > \sqrt{(2g_r r)} \quad \text{(for escape)} \quad (1.12)$$

Substitution of appropriate values of r and g_r shows that the limiting escape velocity is about 11 km s^{-1}; it is independent of the particle's mass.

In the region of diffusive equilibrium the velocities of the gas particles have a Maxwell distribution and when they pass into the exosphere some with the greatest velocities can escape; their number depends on the ratio between the escape velocity and the r.m.s. velocity. The important gases exist as atoms, any molecules having been dissociated by solar radiation, and at the exospheric temperature of about 1000 °K their r.m.s. velocities are as shown in table 1. The velocity of hydrogen atoms is sufficiently near the escape velocity for an important number to leave the top of the atmosphere; for helium atoms the number is very much less, but can be important especially at times when the temperature is great. For oxygen the number is never important: it is less important still for heavier particles.

At the greatest heights atomic hydrogen is the main constituent of the neutral atmosphere at all times; it is continually escaping from the top and being replenished by photochemical dissociation of water vapour near the turbopause. Its flow through the atmosphere from the level of production to the base of the exosphere is controlled by diffusion. The precise concentration depends markedly on the temperature, which controls both the escape rate and the rate of diffusion through the other gases.

TABLE 1. *Particle velocities at* 1000 °K

particle	H	He	O
r.m.s. velocity (km s^{-1})	3	1.4	0.7

Helium is one of the end products of the decay of radioactive rocks on the earth. It enters the atmosphere at the bottom and escapes at the top. As with hydrogen the rate at which it flows through the atmosphere is controlled by diffusion, and depends on the temperature.

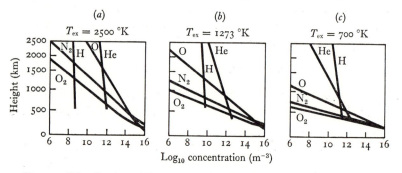

Fig. 1.2. Distribution of the major atmospheric gases at three times when the exospheric temperature (T_{ex}) has different values. (*a*) Day at times of high sunspot number; (*b*) Day at times of medium sunspot number; (*c*) Night at times of low sunspot number. The height scale is adjusted so that the semi-log plots are straight lines in spite of the fact that the scale heights (kT/mg) change as g changes with height (after reference 7).

The temperature at the greatest heights is practically independent of height, it is called the *exospheric temperature* (p. 30). The calculated height-distributions of the important gases are shown, for three different exospheric temperatures, in fig. 1.2. It is noticeable that, because scale height is proportional to temperature, the concentration at great heights is greater when the temperature is greater, for all gases except hydrogen. For hydrogen, however, the concentration is smaller when the temperature is greater, because then the rate of diffusion upwards, and the rate of escape at the top, are greater.

1.3 The sun's radiation

1.3.1 Photon radiation [35, 89, 102, 107]

Although the sun radiates electromagnetic waves over a wide range of wavelengths, only those in narrow portions of the visible and the radio parts of the spectrum reach the earth's surface. It is the others, absorbed in the air, that interest us here, and for direct knowledge of them we rely on measurements made in rockets and satellites. It is found that in the X-ray and ultraviolet regions of the spectrum (often known as the XUV region) there are strong lines superimposed on an irregular background so that a detailed description of the spectrum is complicated. For the present purpose it is sufficient to smooth the spectral irregularities by taking a running average and to discuss the resulting spectral density of energy flux. Table 2 shows the spectral density in $W\,m^{-2}\,nm^{-1}$ near several wavelengths ranging from 1000 nm in the infrared to 10 nm in the X-ray region, and compares the distribution with that from a black-body source at 6000 °K normalized to equal the observed energies at 500 nm. It is at once clear that, although the shape of the solar spectrum corresponds to a source at 6000 °K in the visible part (400–800 nm) of the spectrum and down to wavelengths near 100 nm, on shorter wavelengths the solar radiation is comparatively much more intense than black-body radiation. The difference arises because the shorter wavelengths come from the hotter corona, at a temperature of about 10^6 °K, and the longer from the cooler photosphere, at a temperature of about 6000 °K.

The greater part of the energy is contained in the range of wavelengths from 200 to 1200 nm, where the average spectral density is of order $1.2\,W\,m^{-2}\,nm^{-1}$; the total energy flux is thus about

$$1.2 \times 10^3\,W\,m^{-2}$$

In the part of the spectrum that interests us here, with wavelengths less than 100 nm, the spectral density is of order $3 \times 10^{-5}\,\mathrm{W\,m^{-2}\,nm^{-1}}$ and the total flux of energy is about $3 \times 10^{-3}\,\mathrm{W\,m^{-2}}$. It is this flux of energy in the shorter wavelengths, less than one part in 10^5 of the total, that is responsible for the phenomena to be discussed in this book.

TABLE 2. *Spectral density of power in solar radiation*

A	100	500	1000	1216	2000	3000	5000	10000
B	10	50	100	121.6	200	300	500	1000
C	10^{-5}	5×10^{-5}	3×10^{-5}	(Ly α)	10^{-2}	0.6	2	0·7
D	10^{-80}	10^{-12}	10^{-4}		0.7	1.5	2	0.5
E		3×10^{-3}		5×10^{-3}		1200		

Line A   Wavelength in Å.
Line B   Wavelength in nm.
Line C   Observed spectral density $(\mathrm{W\,m^{-2}\,nm^{-1}})$.
Line D   Spectral density from black body at $6000\,^{\circ}\mathrm{K}$ normalized to equal observed density at 500 nm.
Line E   Observed total power flux in part of spectrum indicated $(\mathrm{W\,m^{-2}})$.

It is interesting to compare these powers with those available in electric supply systems. The generating stations in England and Wales produce about 35000 MW: if this were spread over an area of $10^5\,\mathrm{km^2}$, roughly the area of the country, the power density would be $3.5 \times 10^{-1}\,\mathrm{W\,m^{-2}}$. The power absorbed from the solar spectrum by the atmosphere over the country is thus about one hundredth of that supplied by the generating stations. The total power in the sun's radiation (including the part that penetrates the atmosphere and reaches the earth's surface) is much greater, about 1000 times that supplied by the generating stations.

Although most of the energy in the visible part of the spectrum comes from sources uniformly distributed over the solar disc, some spectral lines, such as H-alpha and the lines from calcium, are emitted chiefly from smaller areas distributed irregularly. At shorter wavelengths, and particularly in the X-ray part of the spectrum, the energy comes predominantly from small regions of that kind, usually near those that emit the visible hydrogen and calcium lines. X-radiation frequently comes from sources that extend for some distance beyond the edge of the visible disc so that during a total optical eclipse there is no simple relation between the intensity of visible light and the electron content of the ionosphere.

The Lyman-alpha line of hydrogen, with wavelength of 121.6 nm in the ultraviolet, is much stronger than the radiations near it in the

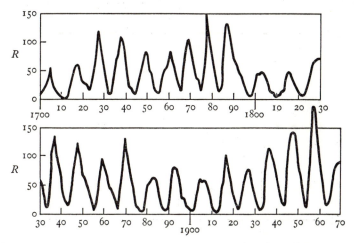

Fig. 1.3. The solar cycle. Sunspot numbers (R) during the period from 1700 to 1970 show a quasi periodicity of about 11 years (based on reference 167).

spectrum; indeed it provides power roughly equal to that in all the rest of the spectrum with wavelengths less than 100 nm, and it produces an important part of the ionization in the low atmosphere (p. 39). Like some of the visible lines it comes from sources irregularly distributed over the solar disc.

Patches of comparative darkness are frequently visible on the sun: they are called *sunspots*. A *sunspot number*, R, based on measurements made by several observatories, is used to represent, in a composite way, the number and area of the spots. The magnitude of R varies between about zero and 200 with a quasi period of about 11 years. This cyclical variation during several past cycles is shown in fig. 1.3: it is called the *solar cycle*. Throughout the solar cycle the energy flux in the ultraviolet part of the spectrum is approximately proportional to $(1 + 0.01R)$; it thus increases by a factor of about 2 as R increases from 0 to 100. In the part of the spectrum with wavelength less than 10 nm (the X-radiation) the changes throughout the cycle are greater: thus as R changes from 0 to 100 the power with wavelengths between 5 and 10 nm increases by a factor of about 3, and that with wavelengths less than 1 nm by a factor of about 30. By contrast the power in the Lyman-alpha line of hydrogen changes by only 50% through the solar cycle, and that in the visible part of the spectrum hardly at all.

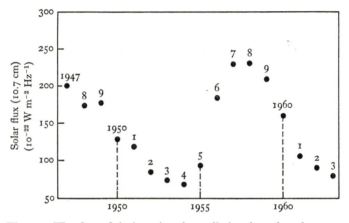

Fig. 1.4. The flux of decimetric solar radiation (wavelength 10.7 cm) varies with the solar cycle (redrawn from reference 7).

Although it is not understood why the sunspot number should provide an approximate measure of the strength of the XUV radiation it has proved so convenient that it has been widely used. In recent times it has been found that the flux of decimetric radio waves (length 10.7 cm) emitted from the sun provides an even better measure. This flux, usually denoted by S, is measured in terms of a unit $10^{-22} \, \text{W m}^{-2} \, \text{Hz}^{-1}$: its variation over a solar cycle is shown in fig. 1.4.

The strength of the XUV radiation often varies rapidly within hours, or days. There is a tendency for large (or small) strengths to repeat after about 27 days, the period of the sun's rotation, as though the radiation were emitted in comparatively narrow beams.

1.3.2 Particle radiation. The solar wind [39, 50, 141]

The solar corona, consisting of fully ionized plasma, mainly protons and electrons, is acted upon by gravity and by internal pressure gradients that depend on the temperature. Because the temperature decreases with radial distance (r) less rapidly than $1/r$ the coronal plasma is accelerated outwards [141] to form a stream of protons and electrons moving with speeds of order 10^5 m s^{-1}. This *solar wind* has been detected by means of satellites at distances from the earth greater than about 15 earth radii. Table 3 lists the magnitudes of some important quantities in the wind in the vicinity of the earth. The power flux in the solar wind is about one-tenth of the flux (3×10^{-3} W m^{-2}) in the XUV spectrum.

TABLE 3. *The solar wind. Quiet sun*

Particle speed (v)	3×10^5 m s^{-1}
Particle flux (F)	1.5×10^{12} m^{-2} s^{-1}
Particle concentration (n) ($= F/v$)	5×10^6 m^{-3}
Energy of a proton in the wind (E)	8×10^{-17} J $= 500$ eV
Energy of an electron in the wind	0.25 eV (negligible compared with proton energy)
Energy density (nE)	4×10^{-10} J m^{-3}
Power flux (nvE)	1.2×10^{-4} W m^{-2}

There are strong magnetic fields on the sun and as the solar wind moves through them currents are induced and the original field is distorted: if the conditions are suitable the distorted field appears to be carried with the wind as though it were 'frozen-in' to the moving plasma (p. 131). The appropriate condition is that the density of kinetic energy in the wind should exceed the density of magnetic energy in the field. At the distance of the earth this field has a strength of about 10^{-9} T (10^{-5} gauss) so that the density of magnetic energy ($B^2/2\mu_0$) is about 4×10^{-13} J m^{-3}: this is much smaller than the density of kinetic energy, so that the particles carry the field with them as they move outwards from the sun.

While the wind is travelling outwards the sun is rotating on its axis once in 27 days so that the particles emitted in succession from a small area are later distributed along a spiral; they carry the magnetic field with them to give it the same spiral shape. At the distance of the earth the radial velocity of the wind, combined with the circumferential velocity appropriate to the 27-day rotation, causes the magnetic field lines to make an angle of about 45° with the radial direction.

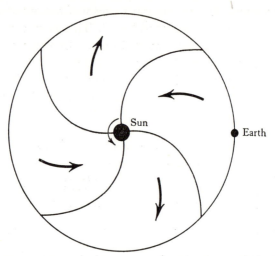

Fig. 1.5. The solar magnetic field carried radially outwards by the solar wind assumes a spiral form because the sun is rotating. Often the field is directed alternately towards and away from the sun as though it was arranged in sectors as shown. The sectors are formed gradually and often persist for several solar rotations.

The spiral magnetic field is often divided into four or six sectors of roughly equal size, such that in neighbouring ones the field is directed in opposite senses as shown in fig. 1.5. The sectors rotate with the sun so that at the earth there are four or six reversals of the field during each solar rotation. The sectors are formed gradually and often persist for many solar rotations.

1.3.3 Solar disturbances [41]

Sometimes disturbances occur on the sun accompanied by the emission of one or more of the following radiations each of which can produce its own type of disturbance in the ionosphere.

1. A sudden increase in the strength of the visible H-alpha line. The phenomenon is known as a *solar flare*.

2. A sudden increase in the strength of the X-radiation: it is known as an *X-ray flare*.

3. The emission of a group of protons and electrons so dense that interaction between the particles gives it the characteristics of a plasma cloud. It forms an enhancement of the normal solar wind and reaches the earth after about 36 hours. It is convenient to call it a *solar plasma event*.

4. The emission of energetic protons and electrons with such small concentrations that they travel like independent charged particles. They reach the earth in a few hours. It is called a *solar proton event*.

X-ray flares, plasma events, and proton events often accompany solar flares. The frequencies of all types of disturbance follow the 11-year solar cycle.

During a flare the enhancement of the X-radiation is greater at the shorter wavelengths; for example, during a flare of importance 2 (on a scale 1 to 3), at an epoch in the solar cycle when $R = 100$, the energy flux in different wavelength ranges exceeds the normal flux by factors (F) roughly as shown in table 4 [35]. In contrast the intensity of the ultraviolet Lyman-alpha line of hydrogen (121.6 nm) does not increase; a surprising fact when it is remembered that the visible flare is itself evidence of a great increase in the intensity of the H-alpha line also emitted by hydrogen.

TABLE 4. *Increased intensity of X-rays during flare of strength 2*

Wavelength range (nm)	0–1	1–2	4.5–6	1–10
Increase by factor F	10	5	4	3

The particles that are emitted during plasma events tend to be ejected in directions roughly normal to the solar surface so that they reach the earth only if their associated sunspots are near the centre of the solar disc; usually the spot has passed the centre by the time the particles have reached the earth. The sun's rotation imposes a 27-day recurrence tendency on those ionospheric phenomena that are associated with the emission of particles in this way.

Near sunspot minimum when there are few visible sunspots, solar plasma events continue to occur fairly frequently: they show well-marked 27-day recurrence tendencies, sometimes lasting over several cycles. It is supposed that the corresponding particles are emitted from invisible isolated regions called *M regions*, that return repeatedly to the side of the sun that faces the earth.

The particles associated with a solar proton event have energies in the range 10^7-10^9 eV. They move as independent particles and are constrained to follow a solar magnetic field line in their travel to the earth. Sometimes, having started on one field line, they are scattered by irregularities in the field on to others and finally reach the earth by a zig-zag route.† The time of their arrival at the earth is not then necessarily related to their energies. Those that come by the more direct routes usually travel from sun to earth in about one hour.

1.4   The action of solar radiations on the atmosphere

1.4.1   Absorption of photon radiation [66]

When radiation from the sun is absorbed in the atmosphere it heats it, dissociates its molecules, and liberates free electrons. The rate at which dissociation or ionization is produced (the rate of production) at any level is proportional to the product of the gas concentration and the intensity of the radiation. At the top of the atmosphere the rate is small because the concentration is small; as the radiation penetrates downwards the concentration increases and with it the rate of production. Below a certain height, however, the strength of the radiation is so much decreased by absorption in the atmosphere that its rate of decrease downwards is greater than the rate of increase of the concentration. There is thus a height where the rate of production reaches a maximum or peak.‡

The approximate level of the peak of production can be deduced from a simple argument as follows. Suppose that the atmosphere consists of molecules (or atoms) all having the same absorption cross-section σ and that radiation falls on it from outside, possibly in an

† If there are appreciable field changes within one gyro radius (about 10^5 km) the particles can pass from one field line to another.

‡ The word 'peak' is used to denote the maximum value of a quantity considered as a function of height.

oblique direction. Then in a column of unit cross-section drawn in the direction of incidence, and of length sufficient to contain N molecules, the total projected absorbing cross-section is $N\sigma$, and when this equals unity the radiation will, on an over-simplified picture, be completely absorbed. The radiation thus descends to a level where the total number of molecules in a (possibly oblique) column of unit area above is $N = \sigma^{-1}$: the peak of production is near that level. This important result is made more precise in appendix A.

When the rate of production is plotted against height the resulting curve represents what has been called a *production layer*. Its shape depends jointly on the natures of the atmospheric gases and of the ionizing radiation. Each gas is distributed in height approximately exponentially and has an absorption (ionization) cross-section that is roughly independent of wavelength over a wide range. It is thus convenient to establish the form of the production layer when ionizing radiation falls (possibly obliquely) on a horizontally stratified gas distributed vertically with a constant scale height and having absorption cross-section independent of wavelength. A layer of that kind was investigated in a classical paper by Chapman [66]: it is now called a *Chapman layer*. The detailed derivation of its shape and behaviour is given in appendix A, where the following important points are established in the equations whose numbers are quoted.

1. The peak of production is at a height where the number of molecules (atoms) in a unit column drawn from that height towards the sun is equal to σ^{-1} (A. 6). This statement is true whatever the height-distribution of the gas.

2. The rate (q) of production is given by

$$q = q_m \exp\{1 - y - \exp(-y)\} \quad \text{(A. 14)} \qquad (1.13)$$

where q_m is the rate at the height (h_m) of peak production, and where $y = (h - h_m)/H$ is the distance above the peak measured in terms of H as a unit.

2.1 The production layers appropriate to any set of parameters can thus all be represented by one curve if heights are measured from the peak in units of H, and production rates are measured as fractions (q/q_m) of the rate at the peak. This curve is the same as that labelled $\chi = 0$ in fig. 1.6.

2.2 The shape depends only on y, that is on H: it is independent of the obliquity (χ) of the radiation, and of the absorption cross-section (σ).

2.3 The absolute magnitude of the peak rate of production per unit volume is

$$q_m = (CI_\infty/eH)\cos\chi \quad \text{(A.9):} \tag{1.14}$$

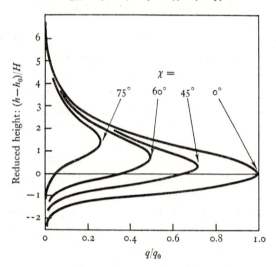

Fig. 1.6. 'Chapman' type production curves for different solar zenith angles (χ). Heights are measured in units of scale height H from the level of the peak when $\chi = 0$. Rates (q) of production are measured in units of q_0, the rate at the peak when $\chi = 0$. If any one of these curves is replotted with normalized co-ordinates described in the text it takes the form of the curve for $\chi = 0$ (after reference 5).

it depends on the intensity I_∞ of the radiation at the top of the atmosphere, on the obliquity (χ) of the radiation, on the efficiency (C) of ionization, and on the scale height (H) of the gas. It is *independent of the ionization cross-section*

3. The height (h_m) of the peak is given by

$$h_m = H\log(\sigma H n_0 \sec\chi) \quad \text{(A.10)} \tag{1.15}$$

it depends on the cross-section (σ), the concentration (n_0) at the height $h = 0$ and on the scale height (H) of the gas, and the obliquity (χ) of the radiation. It is *independent of the intensity I_∞*.

4. If all heights are measured from the height (h_0) of the peak for vertically incident radiation and are expressed in terms of

$$z = (h - h_0)/H,$$

then

$$q = q_0 \exp\{1 - z - \sec\chi \exp(-z)\} \quad \text{(A. 16)} \qquad (1.16)$$

where

$$q_0 = CI_\infty/eH \qquad (1.17)$$

is the peak magnitude of q for $\chi = 0$; fig. 1.6 shows curves for $\chi = 0°$, $45°$, $60°$ and $75°$. All these curves have the shape of that for $\chi = 0°$ if the scales are suitably normalized.

5. Equation (1.13) can be written

$$q = q_m \exp\{1 - \exp(-y)\} \exp(-y)$$

so that at heights that are more than two scale heights above the peak (i.e. $y > 2$ and $\exp(-y) < 0.135$) it is a good approximation to write

$$q \propto \exp(-y) \qquad (1.18)$$

This relation arises, of course, because at those heights the radiation is practically undiminished in strength and the rate of production is proportional to the gas concentration.

1.4.2 Ionization by particle radiation [43]

Protons and electrons, with energies covering a wide range, fall on the atmosphere from outside. Although the ionization they produce is usually small compared with that produced by photons it can be important at times of storm when it is particularly intense, at night when photon radiation is absent, and at small heights which photons cannot reach. The most energetic particles are those of galactic cosmic rays (energy $> 10^9$ eV), next are the particles emitted from the sun during solar proton events (10^7–10^9 eV) followed by the protons of solar plasma events (several keV). At times of storms there are also particles, with energy of several keV that are probably accelerated locally within the magnetosphere.

Charged particles approaching the earth are deviated by the geomagnetic field so that they can reach the lower atmosphere only within limited areas around the poles. Fig. 1.7 shows, on the right-hand scale, the lowest latitudes that can be reached by protons and electrons travelling with the velocities shown on the bottom horizontal scale. The upper horizontal scale gives the times that would be occupied in

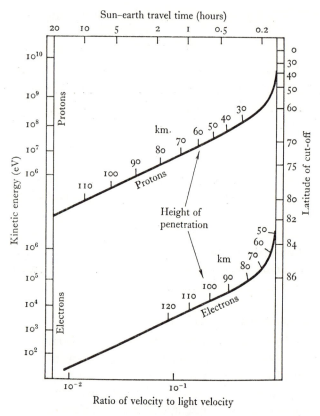

Fig. 1.7. The two curves refer to electrons and protons travelling with velocities as indicated on the bottom horizontal scale. The depths of penetration into the atmosphere are marked on the curves. Reference to the other scales shows (a) the travel times from sun to earth, (b) the kinetic energy, (c) the latitude of 'geomagnetic cut-off' below which the particle cannot reach the earth (after reference 43).

straight line travel from sun to earth, and the left-hand scale shows the energies of the particles. It can be seen, for example, that protons emitted during a solar proton event, reaching the earth after straight travel lasting 1–3 hours, can reach latitudes greater than about 72° (for travel time one hour) or 77° (for travel time three hours) whereas electrons travelling with them are confined to latitudes even greater than 86°. Even galactic cosmic rays with energies greater than 10^9 eV are unable to reach the lowest latitudes.

All charged particles ionize most rapidly near the ends of their paths, when the time spent near an atom is comparable with the period of an electron in a Bohr orbit. When particles are incident on an atmosphere with an exponential height-distribution the rate at which electrons are produced is even more strongly concentrated near the ends of the paths: the heights at which protons and electrons of different energies produce electrons most copiously are shown on the curves of fig. 1.7. Protons that traverse the sun–earth distance in times of 1–3 hours during solar proton events are thus seen to produce electrons most copiously between heights of about 90 and 60 km (the D region). The corresponding electrons ionize most rapidly at heights between about 130 and 100 km, but, as previously mentioned, they are confined to areas very near the poles.

Galactic cosmic radiation is, to some extent, deviated away from the sun's neighbourhood by the solar magnetic field. Since the strength of this field increases with the sunspot number the intensity of the cosmic radiation reaching the earth is weakest at sunspot maximum and strongest at sunspot minimum.

The secondary energetic particles associated with primary galactic cosmic rays penetrate to ground level before reaching the ends of their paths, and they produce electrons at a rate proportional to the concentration of atmospheric particles. The rate is shown, for sunspot maximum and sunspot minimum, in fig. 2.3 (p. 40).

1.4.3   The distribution of electrons or dissociation products after loss and movement

It is next necessary to consider how the electrons, or the dissociation products, formed by solar radiation are distributed in height. After having been produced they may disappear by chemical or ionic reactions, or they may move to other places, so that the shape of their final distribution may differ considerably from that of the production layer. Calculations of these distributions, for different dissociation products, are discussed at many places in this book. Some general principles that apply equally to electrons and to the products of chemical dissociation are established here: for brevity, however, only electrons, produced by ionizing radiation, will be mentioned.

The electrons, produced in unit volume by the radiation, may leave that volume either by being destroyed inside it, or by moving out of it. The number destroyed in unit volume in unit time is usually referred to as the rate of loss, it depends on the concentration (n) of the electrons and can be written $L(n)$.

If the electrons have a common drift velocity $\mathbf{V}$ in addition to their random gas-kinetic velocities, then the number moving out of unit volume in unit time is

$$\operatorname{div}(n\mathbf{V}) = n.\operatorname{div}\mathbf{V} + \mathbf{V}.\operatorname{grad} n \qquad (1.19)$$

There can thus be a change of the number per unit volume because there is either a change of velocity $(\operatorname{div}\mathbf{V})$ or a change of concentration $(\operatorname{grad} n)$ across it. Because $\operatorname{grad} n$ is nearly in the vertical direction it is the vertical component of $\mathbf{V}$ that is important in determining the magnitude of $\operatorname{div}(n\mathbf{V})$. If we denote the upwards vertical velocity by W the rate of removal of electrons from unit volume becomes

$$-\frac{\mathrm{d}n}{\mathrm{d}t} = \frac{\mathrm{d}}{\mathrm{d}h}(nW) = n\frac{\mathrm{d}W}{\mathrm{d}h} + W\frac{\mathrm{d}n}{\mathrm{d}h} \qquad (1.20)$$

The movements to be considered may be the result of diffusion, or of electric and magnetic fields, or of winds blowing in the surrounding gases.

The electrons produced continuously in each unit volume at the rate q are removed by the loss processes and by movement, with the result that the concentration increases at a rate given by the continuity equation

$$\frac{\mathrm{d}n}{\mathrm{d}t} = q - L(n) - \frac{\mathrm{d}}{\mathrm{d}h}(nW) \qquad (1.21)$$

First consider what happens if no new electrons are being produced $(q = 0)$ (for example at night if solar radiation is the only source of electrons) and if there are no movements $(W = 0)$. The electron concentration then decays as given by

$$\mathrm{d}n/\mathrm{d}t = -L(n) \qquad (1.22)$$

In some situations the rate of loss is proportional to the concentration, so that $L(n) = \beta n$: the decay then proceeds so that

$$n = n_0 e^{-\beta t} \tag{1.23}$$

with a time constant β^{-1}. In other situations the electrons are lost by recombining with positive ions so that $L(n) = \alpha n(e)\,n(i)$ where $n(e)$ and $n(i)$ are the concentrations of electrons and ions and α is a recombination constant. Under most of the circumstances to be considered in this book the two concentrations are equal so that we may write $n = n(e) = n(i)$ and $L(n) = \alpha n^2$. The decay then proceeds according to the equation

$$dn/dt = -\alpha n^2 \tag{1.24}$$

If the concentration is n_0 when $t = 0$ its magnitude (n) at time t is given by

$$\frac{1}{n} - \frac{1}{n_0} = \alpha t. \tag{1.25}$$

Although this time-variation is not exponential it is convenient to describe the rate of change at any particular instant as though it were. For this purpose write

$$dn/dt = -(\alpha n)\,n = -\beta n$$

to show that, for a particular magnitude of n, the instantaneous rate of decay is what it would be if the decay were exponential with a time constant $\beta^{-1} = (\alpha n)^{-1}$. This quantity has been called the equivalent time constant, or sometimes the *sluggishness* of the ionosphere.

During the day, as the sun rises and sets, the rate of production by the solar radiation at any level increases to a maximum and then decreases: at some stage the resulting electron concentration also reaches a maximum, so that $dn/dt = 0$ and

$$0 = q - L(n) - d(nW)/dh \tag{1.26}$$

Although this expression applies strictly only when $dn/dt = 0$, it is approximately valid whenever the rate of change (dn/dt) is small compared with the other terms in (1.21). Except near sunrise and sunset the approximation is usually justified simultaneously at all heights and (1.26) can be used to discuss the situation during the daytime. A distribution deduced in this way is said to be a *quasi-equilibrium distribution*.

In different parts of a quasi-equilibrium distribution, one or other of the negative terms in (1.26) frequently predominates. If

$$L(n) \gg \mathrm{d}(nW)/\mathrm{d}h$$

so that $q \fallingdotseq L(n)$ the rate, q, of production by photons is balanced by the loss processes which are chemical in origin: it is then said that there is *photochemical equilibrium*. If $L(n) \ll \mathrm{d}(nW)/\mathrm{d}h$ the production is balanced by the electrons' drifting away from the place where they are produced, it is then said that there is *drift equilibrium*.

In deciding whether any given situation corresponds most nearly to photochemical or to drift equilibrium it is useful to consider how far an electron moves during its lifetime. If, after being produced, and before being destroyed, it moves to a place where the concentration is appreciably different, then drift plays the important part in determining the final distribution; but if the electron is destroyed before it moves so far, loss plays the important part. This idea can be made a little more precise as follows.

It frequently happens that the important part of the movement term in (1.20) arises from the gradient of n rather than of W, so that $W \cdot \mathrm{d}n/\mathrm{d}h$ represents the rate of loss by drift: then the electrons are in photochemical or in drift equilibrium according as

$$L(n) \gg \quad \text{or} \quad \ll W \cdot \mathrm{d}n/\mathrm{d}h \tag{1.27}$$

If the (local) height distribution of n is described in terms of a distribution height δ so that $\mathrm{d}n/\mathrm{d}h = -n/\delta$ and if $L(n) = \beta n$, (1.27) becomes

$$\beta^{-1} \ll \quad \text{or} \quad \gg \delta/W \tag{1.28}$$

Expression (1.28) represents the condition that the relaxation time (β^{-1}) is much less or much greater than the time taken by an electron to drift through a distance equal to one distribution height, a condition previously arrived at by simple reasoning.

Sometimes the movement is the result of diffusion, and then it is convenient to write $\mathrm{d}(nW)/\mathrm{d}h = \gamma n$ where γ^{-1} is the appropriate time constant associated with the movement. It is shown in appendix B that if the electrons are distributed with concentration proportional to $\exp(-h/\delta)$ throughout a gas with distribution height H, then γ is of order $D/(H^2 \text{ or } \delta^2)$ where D is the diffusion coefficient. The distribution is thus approximately

$$\left. \begin{array}{l} \text{in photochemical equilibrium if } \beta > D/(H^2 \text{ or } \delta^2) \\ \text{or in diffusion equilibrium if } \quad \beta < D/(H^2 \text{ or } \delta^2) \end{array} \right\} \quad (1.29)$$

1.5 Chemical effects of solar radiation [12, 48]

1.5.1 Oxygen [125, 129]

It is now necessary to discuss the height-distribution of the oxygen atoms that result from the dissociation of molecular oxygen by radiation with wavelength less than 175 nm. In several other places in this book the ionization (or dissociation) of a parent gas by solar radiation is discussed, and in all of them the radiation is so weak that the concentration of the ionization (or dissociation) product is small compared with the concentration of the parent gas: the calculation can then proceed on the assumption that the distribution of the parent remains unaltered. A different situation is encountered with oxygen. The solar radiation is so intense that the distribution of the molecules is altered to an important extent when they are dissociated and the calculation is no longer simple. The rate of photo-dissociation of molecular oxygen at different heights has, nevertheless, been calculated by several different methods: they all lead to the conclusion that it has a peak of magnitude about $10^{11} \text{ m}^{-3} \text{s}^{-1}$ somewhere in the neighbourhood of 90–95 km. The estimate of this rate depends, of course, on the assumed intensity of the dissociating radiation.

Once the oxygen atoms have been formed they may recombine to produce molecules and they may move vertically under the forces of diffusion. Recombination represented by $O + O \rightarrow O_2$ occurs only slowly, because it is difficult to conserve energy and momentum in a reaction that yields only one particle. The reaction is more rapid when it occurs in the neighbourhood of a third particle M (as represented by $O + O + M \rightarrow O_2 + M$) capable of accepting some of the energy and momentum. The recombination rate is then proportional to $[M] [O]^2$†

† A symbol in square brackets represents the concentration (number per unit volume) of the particle represented by the symbol.

and to the number of collisions made in unit time: at heights near 95 km (where $[M] \fallingdotseq 10^{19} \, \mathrm{m}^{-3}$) the rate can be written

$$d[O]/dt = -\alpha[O]^2$$

with $\alpha = 10^{-25} \, \mathrm{m}^3 \, \mathrm{s}^{-1}$. If photochemical equilibrium were established between the rate (q) of production and the rate of loss then the concentration $[O]$ would be given by $q = \alpha[O]^2$ and $[O]$ would be about $10^{18} \, \mathrm{m}^{-3}$ at the level of the production peak: as stated previously this is comparable to the concentration $[O_2]$ of the original molecular oxygen at that level.

The time constant for the disappearance of the atomic oxygen by recombination to form molecules is of order $1/\alpha[O] = 10^7 \, \mathrm{s}$ or about 100 days. If there were no movements of the atoms and molecules and if the radiation were applied suddenly and remained constant, equilibrium would be established after a time of that order. At greater heights a large proportion of the oxygen molecules would be converted to atoms, which would have a height-distribution like that of the original molecules; meanwhile the molecules would have been largely used up so that the distributions would be as represented by the dashed lines bc' and bd in fig. 1.8.

If the atomic and molecular oxygen, unacted upon by the dissociating radiation, were in stable equilibrium under gravity they would have concentrations represented by the continuous lines be and bc in the figure : these are quite different from those for photochemical equilibrium. Diffusion therefore occurs and the gases move towards their equilibrium positions. If they move only a small distance in the 100 days required for recombination the photochemical equilibrium will not be much altered, but if they move a great distance, the final distribution will approach more nearly that of diffusive equilibrium.

To decide whether the final distribution is controlled by the photochemical processes or by diffusion it is necessary to compare the time constants for the two processes. The gases diffuse, with diffusion coefficient D, through the background of molecular nitrogen, distributed with scale height H: if δ_1 and δ_2 are the distribution heights of the oxygen atoms and molecules the time constant (γ^{-1}) for their diffusion is of order $(H^2 \text{ or } \delta^2)/D$ (appendix B). H is about 10 km and the curves of fig. 1.8 show that $\delta_1 \fallingdotseq 10 \, \mathrm{km}$ and δ_2 is much smaller.

D (equal to $v/\sigma n$ where v is the gas-kinetic velocity, σ the collision cross-section, and n the concentration of the background nitrogen molecules) is about 3×10^3 m^{2}s^{-1} at 90 km so that $\gamma^{-1} \doteqdot H^2/D = 3 \times 10^4$ s. This time constant is much less than the recombination time constant (10^7 s) so that the constituents move a long way towards their diffusive

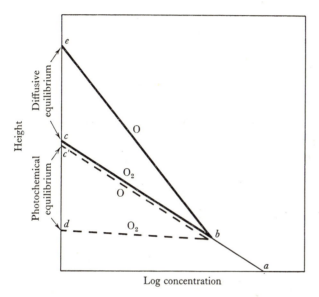

Fig. 1.8. The line *abc* represents the height-distribution of O_2 in the absence of dissociating radiation. If most of the molecules were dissociated at heights greater than *b*, and if there were no movements, the situation would be one of photochemical equilibrium and the dashed lines *bc'* and *bd* would represent the distributions of O and O_2. The concentrations are, however, changed more rapidly by diffusion than by photochemical processes: the two gases therefore move until their distributions become those appropriate to diffusive equilibrium, as shown at *be* (for O) and *bc* (for O_2).

equilibrium positions before recombination or dissociation occurs: the atoms and molecules are thus distributed approximately each with its proper scale height. The atoms produced by the dissociation are continuously falling downwards to below the peak of production where they recombine, while the molecules are continuously diffusing upwards to be dissociated at greater heights. The relative numbers of atoms and molecules is difficult to calculate: experiment shows it to be unity near a height of 120 km. Above that height oxygen is present in both the atomic and molecular forms, each distributed with its own scale height. Distributions at different times of day and different epochs in the solar cycle are shown in fig. 1.2.

1.5.2 Nitric oxide [124]

Nitric oxide is an important minor constituent in the lowest part of the ionosphere because it can be ionized by the intense solar Lyman-alpha radiation. One explanation for its presence suggests that it is formed when atomic nitrogen reacts with molecular oxygen. The necessary nitrogen is not produced in sufficient quantity by photo-dissociation of molecular nitrogen: it is produced at heights greater than 90 km by the action of charged particles, and then diffuses downwards. The responsible reactions are probably (a) the charge-exchange reaction $O^+ + N_2 \rightarrow NO^+ + N$ and (b) the dissociative recombination

$$NO^+ + e \rightarrow N + O \quad (\text{see } \S 2.3).$$

The atomic nitrogen formed in this way takes part in a reaction $N + O_2 \rightarrow NO + O$ that produces NO, and also in a reaction

$$NO + N \rightarrow N_2 + O$$

that destroys it. If the reaction rates for these two processes are k_1 and k_2 and if there is a balance between production and loss of NO, we can write $k_1[N][O_2] = k_2[NO][N]$ or $[NO]/[O_2] = k_1/k_2$. At heights where there is enough atomic nitrogen to make these reactions the most important ones, the ratio $[NO]/[O_2]$ is thus independent of the concentration $[N]$. These heights are probably above 90 km: insertion of the magnitudes of k_1 and k_2 then leads to the ratio $[NO]/[O_2] \doteqdot 10^{-8}$. Roughly the same ratio is maintained at smaller heights by turbulent mixing of the constituent gases. Although only this very small fraction (10^{-8}) of the atmosphere at these levels consists of NO it can play a large part in the production of the electrons in the ionospheric D region ($\S 2.6$).

1.6   The heating of the atmosphere [99]

1.6.1   The temperature of the atmosphere [7, 173]

The energy absorbed from the solar radiation produces ionization, dissociation and excitation of the atmospheric gases: the electrons and ions then recombine, the products of dissociation re-associate, and the excited atoms become de-excited: finally the energy of the particles becomes so small that they can no longer produce excited atoms or molecules. This gas-kinetic energy is ultimately lost by heat conduction to lower levels.

To establish orders of magnitude, consider first what happens to the XUV energy that is absorbed in the ionosphere from the solar radiation. The power Q falling on a column of unit cross-section is about $3 \times 10^{-3}\,\mathrm{W\,m^{-2}}$ and is mainly absorbed at heights greater than about $100\,\mathrm{km}$. This column contains a number (N) of atmospheric particles equal to σ^{-1} where σ is the absorption cross-section, equal to about $10^{-21}\,\mathrm{m^2}$. Each particle has a specific heat C_V of order k (Boltzman's constant) so that if the gas had no means of losing heat it would heat up approximately at the rate

$$\mathrm{d}T/\mathrm{d}t = Q/NC_V = (3 \times 10^{-3})/(10^{21})\,(10^{-23}) = 0.3\,\mathrm{degK\,s^{-1}}$$

or about $1000\,\mathrm{degK\,h^{-1}}$.

At distances more than one scale height above the peak of power input the power absorbed in a layer of gas is proportional to the number of particles in the gas, so that Q/N, and with it the rate of increase of temperature, would be the same at all heights if the heat were not, somehow, removed.

In the real atmosphere the heat is conducted to lower heights and, if the heat input remained constant, the temperature would increase until the rate at which heat was lost by conduction down a temperature gradient ($\mathrm{d}T/\mathrm{d}h$) balanced the rate (Q) at which it was supplied. Orders of magnitude can be calculated from the expression $Q = K\,\mathrm{d}T/\mathrm{d}h$, where K is the heat conductivity; for oxygen it is about

$$3.3 \times 10^{-3} \times T^{\frac{1}{2}}\,\mathrm{J\,m^{-1}\,degK^{-1}}.$$

Then with $Q = 3 \times 10^{-3}\,\mathrm{W\,m^{-2}}$ and $T = 1600\,\mathrm{degK}$ we have

$$\mathrm{d}T/\mathrm{d}h = Q/K = 25\,\mathrm{degK\,km^{-1}}.$$

Below the peak of power input, and under equilibrium conditions, there must be a temperature gradient of this order to conduct the heat downwards.

At any level above the peak of power input the temperature gradient must be sufficient to conduct downwards the power supplied above, and since the input decreases upwards with the gas density whereas the conductivity remains constant,† the required temperature gradient rapidly becomes smaller at greater heights. Above about 250 km the

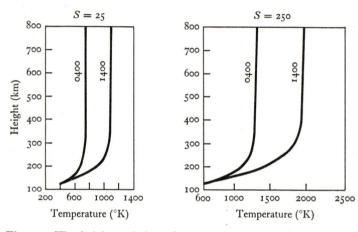

Fig. 1.9. The height-variation of temperature at two times of day and at epochs of the solar cycle when the decimetric flux (S) = 25 and 250 (in units of 10^{-22} W m^{-2} Hz^{-1}) (after reference 7).

temperature is thus approximately independent of height; it is called the *exospheric temperature*. As the solar cycle advances from minimum to maximum the input power increases and with it the temperature gradient near 200 km: this increase is accompanied by an increase in the exospheric temperature. The height-distributions of temperature at two times of day and two epochs of the solar cycle are shown in fig. 1.9.

The heat that is conducted downwards through the temperature

† Simple kinetic theory shows that the conductivity is independent of pressure, since a decrease in the number of molecules available for conduction is balanced by an increase in their 'range of action' measured by their mean free path.

gradient near 200 km is ultimately radiated away, into space or into the earth, by atoms and molecules in the lower atmosphere capable of being excited to radiate in the infrared part of the spectrum.

The time taken for the atmosphere to cool down when the heat source has been removed can be calculated roughly as follows. Suppose that a column of atmosphere with unit cross-section containing heat energy U loses heat by conduction downwards at a rate

$$-\frac{dU}{dt} = aU \qquad (1.30)$$

so that U, and hence the temperature, decreases with a time constant a^{-1}. When the heat source is active, supplying heat at rate Q, the heat energy in the column builds up until the rate of loss by conduction is equal to the rate of supply and

$$Q = aU \qquad (1.31)$$

If there are N particles, with heat capacity Nk, at a temperature T in the unit column, then $U = NkT$. Insertion of the values $N = 10^{21}\,\text{m}^{-2}$, $k = 10^{-23}\,\text{J deg}^{-1}$, $T = 1600\,°\text{K}$ shows that $U = 16\,\text{J m}^{-2}$. Then with $Q = 3 \times 10^{-3}\,\text{W m}^{-2}$ the time constant $a^{-1} = U/Q$ is about $5 \times 10^3\,\text{s}$: the temperature gradients in the upper atmosphere therefore change fairly quickly after sunset.

1.6.2 Atmospheric density deduced from satellite orbits
[110, 127, 146]

Although the density of the atmosphere is very small at heights where most artificial satellites move, it is, nevertheless, often great enough to impede their motion slightly and to alter their orbits after a sufficiently long time. Observations of satellite orbits have thus led to conclusions that verify, and extend, the theoretical results of the previous section.

A satellite with cross-section A encircling the earth with velocity V at a height where the air density is ρ imparts a velocity of order V to a mass of air $A\rho V$ in each second; the rate of change of momentum, equal to the drag force on the satellite, is thus $A\rho V^2$ and the rate at which the satellite loses energy is $A\rho V^3$.

To calculate orders of magnitude suppose that the radius r of the satellite's orbit is equal to the earth's radius (6000 km) so that the velocity $V(= \sqrt{gr})$ is about $8 \, \mathrm{km \, s^{-1}}$; suppose also that a balloon satellite, like Echo I, has a mass $50 \, \mathrm{kg}$, a cross-section (A) $500 \, \mathrm{m^2}$, and that it is at a height of $500 \, \mathrm{km}$ where $\rho = 10^{-12} \, \mathrm{kg \, m^{-3}}$. Then its kinetic energy is $1.6 \times 10^9 \, \mathrm{J}$, the force on it is $3.2 \times 10^{-2} \, \mathrm{N}$, and it loses energy at a rate of about $256 \, \mathrm{J \, s^{-1}}$, as though it were acted upon by a brake dissipating $256 \, \mathrm{W}$. If the rate of loss of energy remained constant (it is actually proportional to V^3) the original kinetic energy would be lost in about $6.5 \times 10^6 \, \mathrm{s}$ (65 days): clearly the air drag is important and observations extending over several days can be used to estimate its magnitude.

Although it might seem that the speed of the satellite should be decreased by the air drag it is in fact increased: this apparent paradox is explained as follows when account is taken of the potential energy (W_p) in the earth's gravitational field. Since W_p is smaller at the smaller radial distances it is convenient to take it as zero when the distance is infinite, then, if G is the gravitational constant and M is the mass of the earth

$$W_p = - GmM/r \qquad (1.32)$$

A simple calculation shows that the kinetic energy (W_k) of the satellite in its circular orbit is given by

$$W_k = GmM/2r \qquad (1.33)$$

Hence the total energy (W_t) is

$$W_t = W_p + W_k = - GmM/2r \qquad (1.34)$$

When energy is transferred from the satellite to the air (1.32) (1.33) (1.34) show that

if the total energy *decreases* by ΔW_t

the potential energy *decreases* by $2\Delta W_t$

and the kinetic energy *increases* by $+ \Delta W_t$;

these changes are accompanied by a *decrease* Δr in the radius of the orbit where $\Delta r/2r^2 \, \Delta W_t = GmM$. The situation is well known in relation to the energy changes that occur when an electron in an atom moves from one Bohr orbit to another.

The decrease in the radius of a circular orbit, resulting from air drag, corresponds to a decrease in the orbital period, so that, if this period is measured, the drag, and hence the air density, can be calculated. If the orbit is elliptical the air drag is greatest at the point of nearest approach to the earth (perigee), where the air density is greatest, and the density at or near that point can be deduced from the observations. Satellites with perigees in different positions have been used to determine how the atmospheric density varies with height at different

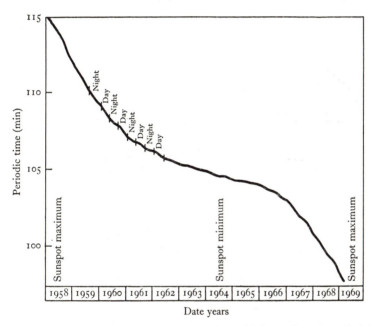

Fig. 1.10. The period of rotation of a satellite with its perigee near a height of 350 km varied more rapidly when the sunspot number was large than when it was small, and during the day than during the night. The air density was greater when the rate of change was more rapid.

places, times of day, and epochs in the solar cycle. Fig. 1.10 shows how the period of a long-lived satellite with perigee at a height near 350 km has changed during ten years. The period is seen to decrease more rapidly during years of high than of low sunspot number and by day than by night. The changes of density deduced from these changes of period are consistent with the changes of temperature shown in fig. 1.9.

There is, however, one experimental result that was unexpected [99]. The density at great heights was found to reach a maximum value soon after mid-day (see fig. 1.11), as though the temperature, also, was then a maximum. The temperature calculated from a knowledge of the intensity of the incident radiation and the thermal conductivity of the atmosphere, with due allowance for changes caused by thermal expansion, was found, however, to differ from the observed temperature in

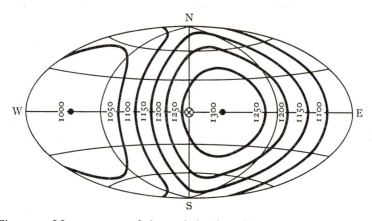

Fig. 1.11. Measurements of the periods of satellite orbits shows that the atmospheric density has a maximum value near 1400 hours. The curves, labelled in °K, show how the exospheric temperature deduced from these results varies over the earth at the equinoxes. The sub-solar point is marked ⊗ (after reference 7).

important ways. Theory indicated that the temperature should reach its maximum at about 1700 hours local time, instead of 1400 hours as observed, and predicted a diurnal range of temperature much greater than was observed. It was therefore suggested that there might be an additional heat source that had a maximum at about 0900 hours.

By using the density determined from satellite orbits in combination with the theory of atmospheric heating and conduction, it is possible to deduce the height-distributions of the atmospheric gases. Those appropriate to a wide range of times of day and epochs in the solar cycle are described in detail in reference number 7: some are shown in fig. 1.2.

In addition to the regular changes of upper atmospheric density

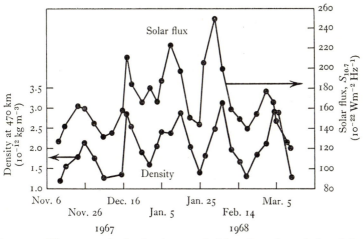

Fig. 1.12. The atmospheric density at a height of 470 km, deduced from measurements on satellite orbits, fluctuates in sympathy with the decimetric flux index S which is itself a measure of the flux of solar XUV radiation.

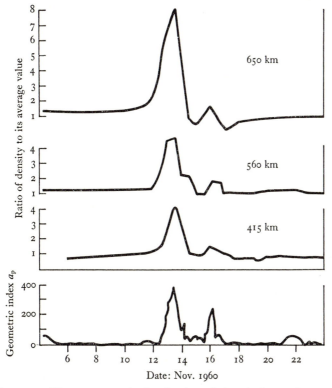

Fig. 1.13. Upper atmospheric density, deduced from observations on satellites with perigees at three different heights, varied in sympathy with the geomagnetic index a_p, which measures the degree of magnetic storminess.

there are irregular, day-to-day, changes accompanied by changes in the decimetric solar flux (see fig. 1.12). The relation between the two presumably arises because the decimetric flux is a good measure of the flux of XUV radiation that heats the atmosphere. Other changes of upper atmospheric density can be related to the occurrence of magnetic and ionospheric storms, as though the atmosphere is heated on those occasions (see fig. 1.13). The mechanism of this heating is not well understood.

2 The ionospheric layers. The sub-peak ionosphere†

2.1 Introduction

The free electrons, produced when the earth's atmosphere is ionized by solar radiation, have a peak concentration at heights near 250 km. It is the purpose of this chapter to discuss the different phenomena that determine their height-distribution, and in particular the distribution at heights below the peak. There is first a discussion of the rates at which electrons are produced at different heights by the ionization of different gases. Next there is a discussion of the rate at which they disappear, either by recombining, or by moving away from their place of origin. The rate of recombination may be so small that the electrons move considerable distances before recombining, or it may be height-dependent: for either, or both, of these reasons electrons may be found in greatest concentration far from the level where they are produced most rapidly.

The electron concentration below the peak has been extensively studied, for many years, with the help of ground-based ionosondes (§9.2). They produce ionograms that emphasize any weakly marked peaks, or even points of inflection, in the height-distribution of electron content: those that occur near heights of 100, 170, and 250 km are said to belong to the E layer, the F1 layer (or ledge) and the F2 layer respectively. Frequently, and always at night, the F1 peak is absent; the single layer above the E layer is then called simply the F layer. The part of the ionosphere below about 90 km requires special methods for its study, and special theories for its explanation, it is called the D region.

From an ionogram it is particularly easy to determine the frequency of a radio wave that just penetrates a layer at vertical incidence and from it to deduce the electron concentration $[e]_m$ at the peak. If the electron distribution represented a state of equilibrium between production by solar radiation incident at an angle χ, and loss by

† [148].

recombination independent of height, then the theory of a simple layer, discussed in appendix A, shows that

$$[e]_m \propto (\cos \chi)^{\frac{1}{2}} \tag{2.1}$$

It is common practice to compare the results of observation with a relation of that kind.

2.2   Rate of production of electrons [35, 106, 132]

In any one gas, ionized by vertically incident radiation, electrons are produced most rapidly at a height where the number of molecules in a superincumbent unit column is equal to the reciprocal of the ionization cross-section. This cross-section varies so gradually with wavelength over wide parts of the spectrum that it can be taken as constant in an approximate calculation. If the concentrations of different ionizable gases are known as functions of height it is then possible to deduce the heights at which different wavelengths would produce electrons most rapidly if they were incident vertically. The curve at *A* in fig. 2.1 shows these heights; the vertical distributions of the important gases are indicated at *B* and the wavelengths that can ionize them are indicated on the horizontal scale at *A*. The figure also indicates the heights occupied by the regions of the ionosphere known by the names D (below 90 km), E (between 90 and 130 km) and F (above 130 km).

The spectrum can be divided roughly into three parts with wavelengths (*a*) less than 14 nm; (*b*) between 14 and 80 nm; and (*c*) between 80 and 102.7 nm, the ionization threshold for O_2. Electrons are produced most copiously in the E region by radiations (*a*) and (*c*) ionizing N_2, O_2, and O; and in the F region by radiations (*b*) ionizing N_2 and O. The rates of production by these three spectral regions are plotted, as functions of height, in fig. 2.2, where the total rate of production, obtained by adding the three, is also shown.

The parts of the spectrum that penetrate to the D region have wavelengths less than about 1 nm (X-radiation) or greater than 102.7 nm. The X-radiation ionizes all the gases and normally produces electrons at the rates indicated on the curve labelled 'X-rays quiet sun' in fig. 2.3. The enhanced X-radiation during a solar flare leads to an increased rate of production, as shown by the curve labelled 'X-rays solar flare'.

Most of the longer-wave radiation that could ionize gases in the D region is absorbed at greater heights. It happens by chance however, that the strong Lyman-alpha radiation with wavelength 121.6 nm

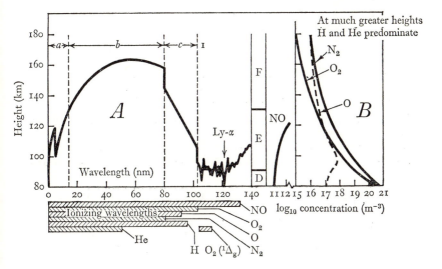

Fig. 2.1. (*A*) Shows the height to which different wavelengths penetrate before being attenuated by a factor e^{-1}. The wavelengths that can ionize the different gases are shown below the horizontal scale. On wavelengths less than 102.7 nm, ionization of at least one of the major gases is the main cause of absorption, the curve thus represents the heights at which electrons are produced most rapidly. The parts of the spectrum marked (*a*) and (*c*) ionize the D and E regions: the part marked (*b*) ionizes the F region.

On wavelengths greater than 102.7 nm the major gases cannot be ionized and the absorption is caused by excitation and photochemical reactions. At heights near 80 km, nitric oxide, although forming only about 10^{-9} of the total atmosphere, is ionized by the very intense Lyman-alpha radiation (121.6 nm) which is little absorbed by the other gases.

(*B*) Shows the height-distribution of the major gases and of the minor constituent NO which is important because it can be ionized by wavelengths that penetrate below 90 km. It also shows the names given to regions occupying the height intervals shown.

suffers little absorption and penetrates to heights below 80 km. This relatively strong radiation ionizes the nitric oxide that is present as a minor constituent in the D region and produces electrons at the rate shown in fig. 2.3.

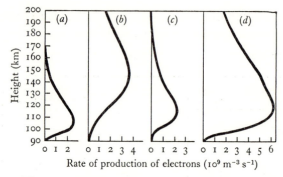

Fig. 2.2. The rates at which electrons are produced at different heights by radiation with wavelengths in the ranges (*a*) less than 14 nm; (*b*) 14–80 nm and (*c*) 80–102.7 nm marked on fig. 2.1. The total rate at which electrons are produced by the whole of the spectrum is shown at (*d*) (after reference 35).

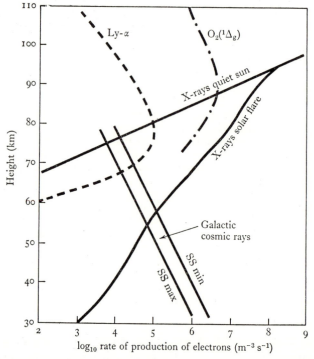

Fig. 2.3. The rates at which electrons are produced in the atmosphere below 110 km. The different curves refer to Lyman-alpha ionizing nitric oxide; X-rays ionizing all gases at times of quiet sun and at times of solar flare; the wavelengths 102.7–111.8 nm ionizing $O_2(^1\Delta_g)$ and galactic cosmic rays at maximum and minimum of the sunspot cycle.

Ionization of excited molecular oxygen ($O_2(^1\Delta_g)$) and ionization by galactic cosmic rays are discussed in §2.6.1.

2.3 The composite F layer [12, 48, 52, 86, 147, 150, 170]

The main peak of electron production is near 150 km and corresponds to the ionization of O and N_2. At greater heights electrons are produced, in each gas separately, at a rate proportional to the concentration of the gas, the two production rates therefore decrease exponentially upwards with distribution heights equal to the scale heights of O and N_2 which are in the ratio 1.75:1.

2.3.1 Recombination

When electrons are produced at the rates discussed above, their equilibrium distribution is determined at the smaller heights by photochemical processes, and at the greater heights by diffusion. If the rate of production is q and the rate of destruction (loss) is L, photochemical equilibrium is set up when $q = L$. A free electron can disappear by reactions of the following three different types:

recombination with an atomic positive ion accompanied by radiation of a photon as represented by

$$e^- + X^+ \to X + h\nu \quad \text{(radiative recombination)} \qquad (a)$$

recombination accompanied by dissociation of a molecular positive ion as represented by

$$e^- + XY^+ \to X + Y \quad \text{(dissociative recombination)} \qquad (b)$$

attachment to a neutral particle to form a negative ion as represented by

$$e^- + Z \to Z^- \quad \text{(attachment)} \qquad (c)$$

At the heights here considered the electrons that become attached by reaction (c) to form negative ions are rapidly detached by other reactions, so that loss of electrons by attachment can be neglected. In the lowest ionosphere the loss of electrons by attachment can, however, be important (§2.6.3).

When electrons disappear by reactions of the types (a) or (b), the rate of loss from unit volume is given by

$$L = -\mathrm{d}[e]/\mathrm{d}t = -\alpha[e][P^+]$$

where α is the recombination coefficient and $[P^+]$ is the concentration of the (atomic or molecular) positive ion. The ionospheric plasma is electrically neutral everywhere (if it were not, electric fields would be set up that would move the charges until neutrality were re-established†); since there are no negative ions we can thus write $[e] = [P^+]$ and $-\mathrm{d}[e]/\mathrm{d}t = -\alpha[e]^2$.

The magnitudes of the recombination coefficients for the two reactions of type (a) and (b) are very different. Because reaction (b) results in two material particles, whereas (a) results in only one (and a photon), the need to conserve energy and momentum is more easily satisfied in (b) than in (a), and the recombination is correspondingly more likely. The calculated magnitude of the radiative recombination coefficient (reaction (a)) is about $10^{-18}\,\mathrm{m^3\,s^{-1}}$ whereas that of the dissociative recombination coefficient (reaction (b)) is very much larger, about $10^{-13}\,\mathrm{m^3\,s^{-1}}$. It thus follows that electrons recombine with molecular nitrogen ions, by a process of the dissociative type (b), comparatively rapidly; whereas they recombine with atomic oxygen ions, by the radiative process (a), comparatively slowly.

It is useful to estimate the rates at which electrons would disappear from the ionosphere, after the ionizing radiation has been removed at night, if they were lost by these reactions. The effective time constant of their disappearance is $(\alpha[e])^{-1}$ (p. 23) and with $[e] = 10^{12}\,\mathrm{m^{-3}}$ (corresponding to the most densely ionized portion of the ionosphere at heights near 250 km) it is of order 10 s for recombination with molecular nitrogen and 10^6 s (300 hours) for recombination with atomic oxygen: with $[e] = 10^{10}\,\mathrm{m^{-3}}$ (corresponding to a height of 1000 km) the two times are 10^3 s and 10^8 s (30 000 hours). The N_2^+ ions thus disappear comparatively rapidly after sunset,‡ but if radiative recom-

† It is shown in §6.2.1 that a very small departure from neutrality does, in fact, occur and provides the electric field that causes the ions and electrons to have the same distribution height. This departure from electrical neutrality is, however, negligible for the present discussion.

‡ It is probable that the majority of the N_2^+ ions disappear by the reaction $N_2^+ + O \rightarrow NO^+ + N$ and that the electrons then recombine with the new

bination were the only mechanism for removing the O^+ ions they would remain, together with the electrons that accompany them, throughout the night. There is, however, another mechanism for their removal as follows.

2.3.2   Reactions involving charge-exchange or ion–atom rearrangement

When a positive oxygen ion collides with a neutral molecule of nitrogen the ion can exchange places with one of the atoms in the molecule by the ion–atom rearrangement

$$O^+ + N_2 \rightarrow NO^+ + N; \qquad (d)$$

when it collides with a neutral oxygen molecule the charge can be exchanged in the reaction

$$O^+ + O_2 \rightarrow O_2^+ + O \qquad (e)$$

Once either of these reactions has taken place there is a positive molecular ion (NO^+ or O_2^+) available, with which an electron can recombine by the comparatively rapid process of dissociative recombination. Those electrons that do not recombine with N_2^+ are removed, together with an equal number of O^+ ions, by this double process. Let us examine the height-distribution of the electrons that result from this type of photochemical equilibrium.

Suppose that reaction (d) is the appropriate one, then under equilibrium conditions the rate at which it removes ions from unit volume is equal to the rate (q) at which they are produced, and

$$q = k_d[O^+][N_2] \qquad (2.2)$$

where k_d is the rate coefficient. The electrons, produced also at the rate q, disappear by the reaction

$$e + NO^+ \rightarrow N + O$$

so that
$$q = \alpha[e][NO^+] \qquad (2.3)$$

molecular ion NO^+ by a dissociative reaction. Because the successive reactions each result in two particles they are both rapid and the effective recombination coefficient has the same order as that for direct recombination with N_2^+.

where α is the appropriate dissociative recombination coefficient. From (2.2) and (2.3) we have

$$[O^+] = q/k_d[N_2] \qquad (2.4)$$

and

$$[NO^+] = q/\alpha[e] \qquad (2.5)$$

Now because the ionosphere is neutral $[e] = [O^+] + [NO^+]$ so that

$$[e] = \frac{q}{k_d[N_2]} + \frac{q}{\alpha[e]} \qquad (2.6)$$

or

$$\frac{1}{q} = \frac{1}{k_d[N_2][e]} + \frac{1}{\alpha[e]^2} \qquad (2.7)$$

In these expressions $k_d[N_2]$ represents a loss coefficient that has the same height-dependence as the nitrogen molecules, it is convenient to write it $\beta(h)$ so that

$$\frac{1}{q} = \frac{1}{\beta(h)[e]} + \frac{1}{\alpha[e]^2} \qquad (2.8)$$

The resulting electron concentration $[e]$ depends on the relative magnitude of the two terms on the right-hand side of (2.8). They are equal at a height, which we shall call the transition level (h_t) for the loss term, given by

$$\beta(h_t) = \alpha[e] \qquad (2.9)$$

Below h_t only the second term in (2.8) is important and

$$q = \alpha[e]^2 \qquad (2.10)$$

Above h_t only the first term is important and

$$q = \beta(h)[e] \qquad (2.11)$$

If (2.10) were applicable at all heights the electron distribution would be as shown in fig. 2.4(*b*) and would peak at the height h_0 where there is a peak of q. If (2.11) were applicable at all heights the distribution would be as shown in fig. 2.4(*d*). Below the peak of production q increases upwards and $\beta(h)$ decreases upwards so that $[e]$ increases. A short distance above the peak, q decreases upwards proportionally to the concentration of the atomic oxygen being ionized, i.e. like $\exp\{-h/H(O)\}$ and $\beta(h)$ decreases upwards proportionally to the concentration of nitrogen molecules, i.e. like $\exp\{-h/H(N_2)\}$. Since the upwards decrease of $\beta(h)$ is more rapid than that of q the electron concentration [e] *increases* upwards like

$$[e] \propto \exp\left[+h\left\{\frac{1}{H(N_2)}-\frac{1}{H(O)}\right\}\right]$$

or
$$[e] \propto \exp\{+0.75h/H(O)\}$$

since
$$H(O)/H(N_2) = 1.75.$$

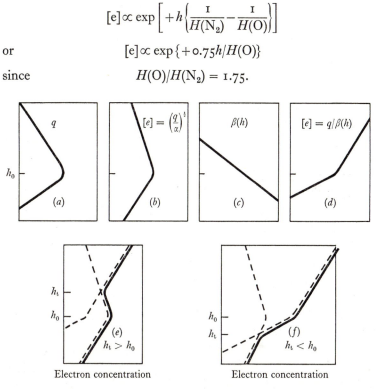

Fig. 2.4. The formation of the F 1 layer. The quantities indicated are plotted on a logarithmic scale against height on a linear scale. (a) shows the rate of production (q) of electrons. The broken lines in (e) and (f) show the curves of (b) and (d) plotted with different positions along the horizontal scale. They cross at the height h_t where $(q/\alpha)^{\frac{1}{2}} = q/\beta(h)$, i.e. where $\beta = \alpha[e]$. The continuous lines represent the sums of the quantities plotted. In (e) $h_t > h_0$ and there is a peak of [e] at h_0: in (f) $h_t < h_0$ and there is no peak of [e].

In the ionosphere there is a transition from the situation of (2.10) (fig. 2.4(b)) to the situation of (2.11) (fig. 2.4(d)) at the height h_t. Fig. 2.4(e) shows the resulting electron distribution when $h_t > h_0$; there is a peak of electron concentration at the height h_0, it is the peak of the F 1 layer. Fig. 2.4(f) shows the distribution when $h_t < h_0$, there is then no peak, and the F1 layer is not observed. The factors that determine the relative magnitudes of h_t and h_0 are discussed in § 2.4.

If either of the conditions of photochemical equilibrium discussed above were to remain appropriate the electron concentration would increase indefinitely as the height increases above the level h_t.

2.3.3   Diffusion [151]

At sufficiently great heights, however, diffusion becomes important and the electron concentration decreases upwards. This diffusion-dominated portion of the plasma, in which the concentration *decreases* upwards, rests upon the lower portion, dominated by photochemical processes, in which the concentration *increases* upwards: where the two meet the concentration reaches a peak. In the diffusion-dominated part, above the peak, it is shown in § 6.2.1 that the distribution height (δ) of the plasma is equal to the scale height of a gas in which the mass of the molecules is the mean of the masses of the ions and the electrons, so that, the electron mass being negligible compared with the mass m_i of the ions, $\delta = 2kT/gm_i$. A measurement of δ can thus often be used to determine the mass (m_i) of the ions.

The height of the peak is determined, approximately, by equating the time constants for the photochemical and for the diffusion processes: it thus occurs where $k_d[N_2] = D/\delta^2$, δ being the distribution height of the plasma, or of the background molecular nitrogen, and D being the diffusion coefficient (p. 107). Insertion of numerical quantities shows that the peak is formed near a height of 300 km; it is the F peak of electron concentration that forms such a noticeable feature of ionograms recorded by ground-based or satellite-borne ionosondes. It is important to realize that it does *not* correspond to a peak in the rate of production of electrons.

2.4   The F1 layer [52, 147]

The F1 peak, or ledge, of electron concentration appears when the height (h_t) of the transition level for the loss coefficient is greater than the height (h_0) of the peak of electron production. The relative magnitudes of h_t and h_0 change with the solar cycle, and with the inclination (χ) of the sun's radiation, so that the likelihood of the F1 layer's being formed changes throughout the cycle and throughout the day and the seasons.

At solar minimum, or when χ is large, [e] has its smallest values; the relation $\beta(h_t) = \alpha[e]$ then requires that β should be small, and since β decreases as height increases, this means that h_t is large. The probability that the F1 peak is noticeable is thus greater at solar minimum than at solar maximum.

The effect of a change in χ is a little more difficult to assess because, although an increase in χ results in an increase in h_t, it also results in an increase of h_0. Detailed calculation shows that the increase of h_0 is greater than the increase of h_t. It follows that $h_t - h_0$ is smallest when χ is greatest, so that the F1 layer is least likely to be observed in winter, and at times far from midday.

When the F1 layer or ledge appears, the electron concentration at its peak is related to the solar zenith angle roughly as represented by (2.1). There are, however, comparatively small systematic departures from this relation which are different at different places and parts of the solar cycle; the forms of the departures also depend on whether the changes of χ correspond to changes in the time of day, the season, or the position on the earth. Attempts have been made to explain these variations in terms of

(a) changes in the transition level (h_t) of the loss process, consequent upon changes in atmospheric composition; or

(b) changes in the magnitudes of the reaction constant k_d supposed to be temperature-dependent.

When values of the intensity (I_∞) of solar radiation (measured by rockets and satellites), and of the peak electron concentration (measured by ionosondes), together with the theoretical value of the coefficient (C) of efficiency of ionization are inserted into the equation (see A 9a):

$$\alpha[e]_m^2 = q_m = (I_\infty C/eH)\cos\chi \qquad (2.12)$$

the value deduced for α at a height near 170 km, is about $3.7 \times 10^{-14} \, m^3 \, s^{-1}$; in reasonable accord with theory.

2.5   The E layer [51]

Below the F1 layer, another layer, the E layer, is always clearly noticeable at a height near 110 km during the day. It is formed by the radiation responsible for a production peak at the same height (see fig. 2.2). Because the peak is below the level h_t the loss-process is one of dissociative recombination, with coefficient nearly independent of height, and the peak electron concentrations measured at different values of χ satisfy (2.1) reasonably well. When values of the peak rate of production, such as those of fig. 2.2 are inserted in (2.12) the value deduced for α is about four times greater than the value for the F1 layer. Since there are theoretical reasons for supposing that α is temperature-dependent, this difference is usually attributed to different temperatures at the two levels.

Although the height of the peak and the peak electron concentration of the E layer vary with the solar zenith angle roughly as for a simple recombination layer, detailed observations have revealed some regular differences: they are, on the whole, more marked than the differences in the corresponding quantities appropriate to the F1 layer. The failure of the E layer to behave like a simple layer has been explained, sometimes in terms of height-variations of the recombination coefficient, and of the scale height, associated with a height-variation of temperature: and sometimes in terms of the electrodynamic forces arising from the atmospheric dynamo currents that flow in it (p. 84).

2.6   The D region [33, 51, 121, 128, 163]

In the D region collisions between electrons, ions and neutral particles are comparatively frequent and ions formed in three-body reactions are important. Rocket-borne mass spectrometers provide evidence about the relative abundance of these ions at different heights. Theories of the region are concerned with the ionizing processes and the reactions that lead to the formation and destruction of the different ions. The situation is complicated and ideas are evolving rapidly while this book is being written. The account given here is intended to illustrate the natures of the processes that are likely to occur; it must not be supposed that the examples given are necessarily the most important.

2.6.1   The ionizing radiations [43]

At heights near 80 km oxygen and nitrogen are ionized by X-radiation of wavelength less than 1 nm, and excited oxygen molecules in the metastable state $O_2(^1\Delta_g)$ are ionized by radiation with wavelengths between 102.7 and 111.8 nm. Nitric oxide is ionized by the relatively intense Lyman-alpha radiation from the sun. Galactic cosmic radiation provides an additional source of ionization, important in the lowest part of the region. The secondary particles produced by the primary radiation are absorbed only slightly even after passing through the whole of the atmosphere, so that the rate at which they ionize is proportional to the concentration of the atmospheric gases, and thus increases steadily downwards. Fig. 2.3 shows the rates at which electrons are produced near sunspot minimum at different heights by the different radiations: it also shows the increased rate of production by X-radiation during an intense solar flare.

As the solar cycle advances, the intensities of the three radiations change in different ways. Between solar minimum and maximum the strength of Lyman-alpha radiation increases by about 50 per cent but that of X-radiation increases by a factor of about 10^3. In contrast galactic cosmic radiation is *weaker* at solar maximum than at solar minimum by a factor of about 0.5: this change is attributed to its being partially deviated away from the neighbourhood of the earth by the solar magnetic field, which is greater at solar maximum. These differences in the rates of ionization cause different parts of the D region to vary throughout the solar cycle in different ways.

2.6.2   Ion chemistry [12, 48, 172]

The ion chemistry of the region is largely controlled by constituents, such as O, O_3, NO, NO_2, CO_2, H_2O, and alkali metals, present only in very small proportions. Because some of these minor constituents last for a comparatively long time before being destroyed, their distribution is largely determined by movements, and since these are related to movements at lower levels the behaviour of the region is, to some extent, influenced by that of the non-ionized atmosphere beneath. Most of the minor constituents are produced or destroyed by solar photon radiation and possibly by particle radiation during storms:

their concentrations depend on the intensities of these radiations, Some, such as alkali metals, are introduced at varying rates by meteors. Because so many phenomena contribute to the variability of the minor constituents in the D region, its behaviour is less regular than that of the ionosphere at greater heights.

In discussions of the ionospheric ions (positive or negative) it is necessary to distinguish between the primary ones and the secondary ones that are produced from them by chemical reactions. The primary positive ions are produced by the action of ionizing radiation on major and minor neutral constituents; the primary negative ions are produced by electron attachment. An ion of one kind can produce another kind by transferring its charge to a neutral particle, if the new ion has an ionization energy less than the old (for positive ions) or if it has a greater electron affinity (for negative ions). When several kinds of ion are formed by successive reactions, one is finally formed with ionization energy so small, or electron affinity so great, that it cannot lose its charge simply by passing it to another particle. A terminating ion of that kind usually disappears by recombination with a charged particle of the opposite sign.

2.6.3   Negative ions [12, 48]

In the lowest ionosphere electrons are removed comparatively rapidly in three-body reactions that result in the formation of negative ions. In one process of this kind electrons become attached to oxygen molecules, in the presence of a third body, to produce negative ions by the reaction

$$e + O_2 + M \rightarrow O_2^- + M \qquad (2.13)$$

which occurs at a rate $k_1[e][O_2][M]$. The resulting O_2^- ion can then react with other constituents to produce other kinds of negative ion. Here the kinds of reaction that can occur are illustrated by supposing that O_2^- is the only negative ion.

This ion subsequently disappears either by 'recombining' with a positive ion or by a process in which the electron is detached to become free again. 'Recombination' with a positive ion occurs in a reaction

$$O_2^- + X^+ \rightarrow O_2 + X \quad \text{(recombination)} \qquad (2.14)$$

with a rate $\alpha_i[O_2^-][X^+]$. Although this process is often called 'ionic recombination', a more correct term would be 'ion–ion neutralization'.

An electron can be detached from O_2^- by any one of the following processes

(a) by collision with another particle (M) as represented by

$$O_2^- + M \rightarrow e + O_2 + M \quad \text{(collision detachment)} \quad (2.15)$$

with a rate $k_2[O_2^-][M]$,

(b) by interaction with atomic oxygen to produce a molecule of ozone and a free electron by the reaction

$$O_2^- + O \rightarrow O_3 + e \quad \text{(associative detachment)} \quad (2.16)$$

with rate $k_3[O][O_2^-]$: or

(c) by the action of photon radiation ($h\nu$) with intensity I as represented by

$$O_2^- + h\nu \rightarrow O_2 + e \quad \text{(photo-detachment)} \quad (2.17)$$

with rate $k_4[O_2^-]I$.

At most times attachment and detachment occur so rapidly that they nearly balance each other, so that the ratio (λ) between the concentrations of negative ions and free electrons can be calculated by equating the rates of production and loss of ions, thus

$$k_1[e][O_2][M] = k_2[O_2^-][M] + k_3[O_2^-][O] + k_4[O_2^-]I$$

Hence
$$\lambda = \frac{[O_2^-]}{[e]} = \frac{k_1[O_2][M]}{k_2[M] + k_3[O] + k_4 I} \quad (2.18)$$

When attachment and detachment of electrons occur it is necessary to modify the continuity equation that relates the concentration of electrons to the rates at which they are produced and lost. It will be supposed that the rates of attachment and detachment are not necessarily equal; and the resultant rate at which the electrons in unit volume are lost by conversion into ions will be represented by L. Then, if it is supposed that electrons are also lost by recombination with positive ions (X^+) with recombination coefficient α_e, the equation of continuity for electrons is

$$d[e]/dt = +q - \alpha_e[e][X^+] - L \quad (2.19)$$

The corresponding continuity equation for negative ions is

$$d[O_2^-]/dt = -\alpha_i[O_2^-][X^+] + L \tag{2.20}$$

The electrical neutrality of the ionosphere requires that

$$[X^+] = [O_2^-] + [e] = (1 + \lambda)[e] \tag{2.21}$$

and if we insert this into (2.19) and (2.20), and add, we obtain

$$(1 + \lambda)\, d[e]/dt = q - (\alpha_e + \lambda\alpha_i)(1 + \lambda)[e]^2 \tag{2.22}$$

or
$$\frac{d[e]}{dt} = \frac{q}{1+\lambda} - (\alpha_e + \lambda\alpha_i)[e]^2 \tag{2.23}$$

In discussions of a quasi-equilibrium situation (where $d[e]/dt \fallingdotseq 0$) it is convenient to write (2.22) in the form

$$q = \psi[e]^2$$

where
$$\psi = (\alpha_e + \lambda\alpha_i)(1 + \lambda) \tag{2.24}$$

In discussions of transitory phenomena, related to sudden changes of q (as in eclipses, or during solar flares) or to decay of electrons at night, it is useful to use the form (2.23) with $q = 0$ and to write

$$\frac{d[e]}{dt} = -\alpha_{eff}[e]^2 \tag{2.25}$$

with
$$\alpha_{eff} = (\alpha_e + \lambda\alpha_i)$$

In the D region the electron–ion recombination coefficient (α_e) has a magnitude about $10^{-13}\,\text{m}^3\,\text{s}^{-1}$, appropriate to dissociative recombination, and the ion–ion recombination coefficient (α_i) is of the same order. In parts of the ionosphere where λ is of order unity or greater the ion–ion 'recombination' is thus significant.

At middle and high latitudes the lowest parts of the D region are ionized chiefly by cosmic radiation which is deviated by the geomagnetic field so that it falls equally on the day and night sides of the earth with intensity that changes little throughout the 24 hours. It might therefore seem that the electron concentration, $[e]$, given by the equilibrium value $[e] = (q/\psi)^{\frac{1}{2}}$, would change little between day and night. But it must be remembered that solar radiation can detach electrons from the negative ions during the day so that then λ (and with it ψ, see (2.24)) is comparatively small, and the electron concentration is comparatively great. It follows that the electron concentration in the lowest ionosphere has a diurnal variation like that of fig. 2.5, it is greater by day than by night and changes abruptly when the detaching radiation reaches, or leaves, the relevant height.

The part of the solar spectrum responsible for the detachment of electrons from negative ions must contain photons with sufficient energy. If the ions are O_2^-, as suggested previously, the detachment energy is 0.43 eV and the corresponding wavelength is 2870 nm so that light in the visible part of the spectrum would suffice. The abrupt

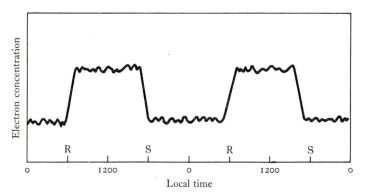

Fig. 2.5. To illustrate how the electron concentration in the lowest part of the D region varies with time of day. R and S represent the times of sunrise and sunset.

changes in the electron concentration near sunrise and sunset would thus be expected to occur when the shadow of the solid earth sweeps across the ionosphere. In fact they are related more nearly to the shadow that would be formed by a surface in the ozone layer at a height about 50 km above the ground.

To explain this observation it has been suggested that photo-detachment is much less important than other processes. If the negative ions were mainly O_2^- one such process would then be associative detachment of the kind described by (2.16). Then, in (2.18) $k_4 I \ll k_3[O] + k_2[M]$ and the change in I at ground sunrise would produce only a small change in λ, and hence in the electron concentration. The sudden change observed when the ozone shadow leaves the

D region is then ascribed to an increase in the concentration of atomic oxygen, and liberation of free electrons by reaction (2.16). The suggestion is that ultraviolet light, no longer absorbed in the ozone layer, dissociates the molecular oxygen, and the resulting atomic oxygen increases the term $k_3[O]$ in (2.18) so as to reduce λ.

Equation (2.18) shows that, since $[O_2]$ and $[M]$ both increase downwards, λ increases downwards also, both by day and by night. Its magnitude has been determined experimentally and theoretically but experiment and theory are difficult, and reliable results are not yet available. It is probable that, by day, $\lambda = 1$ around 70 km and 10 around 60 km: by night the magnitudes are probably greater by a factor of about 10.

3 The F layer peak and above

3.1 The peak of the F 2 layer

The measured magnitude of the peak electron concentration in the F 2 layer is not even approximately related to the angle of arrival of the sun's rays in the way that might be expected from the theory of its formation. Numerous attempts have been made to explain the different kinds of behaviour (sometimes called 'anomalies') observed at different times and places. The causes that have been suggested include changes in the loss rate of electrons consequent upon changes in relative concentrations of atomic oxygen and molecular nitrogen: changes in the reaction constant k_d (2.2) resulting from changes in upper atmospheric temperature: and production of ionization by energetic particles entering from the magnetosphere. It has also been suggested that movements of the ionospheric plasma caused by winds in the neutral atmosphere, or by electric fields originating lower in the ionosphere, might carry the peak to places where the loss coefficient had different values.

It has usually been possible to explain any individual ionospheric peculiarity in terms of one or more of these causes, but it has not yet been firmly demonstrated from independent evidence that the postulated cause does, in fact, operate. There seems little doubt, however, that bodily movements can be responsible for part of the anomalous behaviour of the layer.

3.1.1 Movement of the peak [84, 112, 171, 174]

Ionospheric ions and electrons can be moved by forces arising from winds in the surrounding neutral gas (air drag) or from electric fields transmitted from another part of the ionosphere. At heights where the collision frequency (ν_i) of the ions is much less than their gyrofrequency (Ω_i) mechanical forces, imposed by a wind in the neutral air, move the electron–ion plasma most readily along the direction of the magnetic field (p. 112). In this way winds, driven by the afternoon bulge of atmospheric pressure and influenced by Coriolis forces, could cause the ionospheric plasma to move along the sloping lines of the geomagnetic field. If the movement is downwards it is to a place where electrons are lost more rapidly, whereas if it is upwards they are lost more slowly. It has been suggested that these changes in the rates of loss can explain some of the less regular features of the F layer at heights greater than about 140 km where $\nu_i \ll \Omega_i$.

At these same heights the electric field that has its origin in the atmospheric dynamo near a height of 110 km causes the ions and electrons to move together in a direction perpendicular to the magnetic field as though they formed the armature of an atmospheric motor (p. 84). The vertical component of this movement can be important in determining the height of the F layer peak, and hence the rate of loss of electrons.

3.1.2   Behaviour near the geomagnetic equator [77, 93]

Near the geomagnetic equator, where the magnetic field is horizontal, the movement resulting from an imposed E–W electric field is vertical: it is upwards during the day and, when combined with preferential

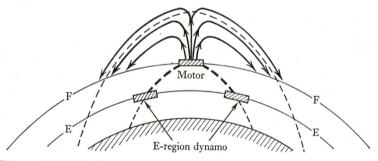

Fig. 3.1. The F region geomagnetic anomaly. Near the equator the electric fields of the atmospheric dynamo in the E layer are conveyed upwards along geomagnetic lines of force to the 'motor' in the F layer where they produce an upwards movement of the plasma during the day. The raised plasma then diffuses down lines of force to produce enhanced concentration at places on each side of the equator, and decreased concentration at the equator itself.

diffusion along the direction of the magnetic field, leads to an interesting phenomenon that has been called the *geomagnetic anomaly* (fig. 3.1). After the ionospheric plasma has been moved upwards it diffuses downwards again, but obliquely along the lines of force, so that instead

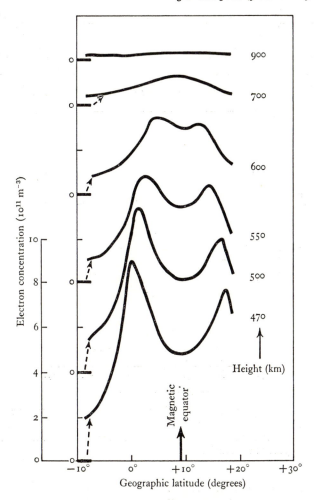

Fig. 3.2. In the daytime topside ionosphere the electron concentration is increased along a geomagnetic line of force that reaches up to a height of about 700 km. When the concentration is plotted, at lower heights, against latitude there is thus a minimum on the magnetic equator and maximum on each side. The maxima are closer together at the greater heights.

of returning to its source it arrives at two places, one to the north and one to the south. The electron concentration is thus depleted on the geomagnetic equator and enhanced in two regions, one on each side. Topside sounder experiments show this distribution in the form of an arch in the upper part of the layer (fig. 3.2) which changes its shape, through the day, and during ionospheric storms, in the way to be expected from theories of the atmospheric dynamo and the electric fields produced by it. The arch of increased concentration is also observable below the peak of the F layer as shown in fig. 3.3.

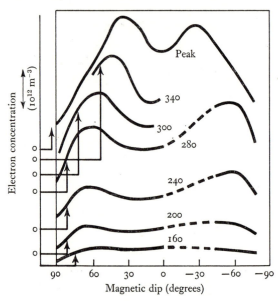

Fig. 3.3. At, and below, the peak of the F layer the electron concentration at any one height is a minimum on the magnetic dip equator, and is maximum at places north and south. The maxima fall on a geomagnetic field line and are farther apart at smaller heights.

3.2   Above the F layer peak [31, 49, 61, 100]

3.2.1   Oxygen ions

A short distance above the F layer peak the effects of diffusion become much greater than those of electron production or loss and the concentration of the ion–electron plasma assumes the height-distribution appropriate to diffusive equilibrium under gravity. If the electrons and positive ions were to move independently their very different masses would result in their assuming very different height-distributions: electrostatic forces, however, compel their concentrations to be the same with a distribution height equal to twice the scale height of the neutral gas from which the ions are derived (p. 107). the electron concentration [e] a short distance above the F layer peak is thus given by

$$[e] = [e]_0 \exp\{-h/2H(O)\} \tag{3.1}$$

where $H(O)$ is the scale height of atomic oxygen.

At the greater heights free diffusion can occur only along the direction of the geomagnetic field so that (3.1) describes the distribution along a field line. For example, at heights represented by h_1, h_2, h_3, h_4 on fig. 3.4 it relates the magnitude of [e] at places such as a, b, c, d but not

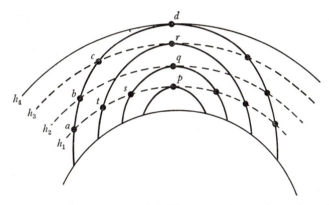

Fig. 3.4. The relation $n = n_0 \exp(-h/H)$, appropriate to charged particles in diffusive equilibrium, applies at points such as a, b, c, d along a geomagnetic field line, but not at points such as p, q, r, d along a vertical. The distribution p, q, r, d along a vertical at the geomagnetic equator is related to the (latitudinal) distribution p, s, t, a along a horizontal.

at places p, q, r, d. The figure also shows that the vertical distribution at the places p, q, r and d above the geomagnetic equator is related, through (3.1), to the latitudinal distribution at the places p, s, t, a, at a constant height, and in particular at the peak of the F layer. The observed sharp gradient in the radial distribution at the equator (known as the plasmapause) thus corresponds to a sharp gradient in the latitudinal distribution at the peak of the F layer (known as the mid-latitude trough in the F region) (p. 76).

3.2.2 Hydrogen ions

In the highest part of the atmosphere atomic hydrogen is the predominant gas and sometimes, between it and the atomic oxygen below, there is a region where helium predominates (fig. 1.2). At great heights it is thus necessary to consider the rates of production of electrons from hydrogen and helium. For each of these gases the heights to be considered are such that $N \ll \sigma^{-1}$, so that they are considerably above the peaks of production: the ionizing radiation is thus little absorbed, and the rates of production are proportional to the concentrations of the gases.

Hydrogen atoms are ionized by solar ultraviolet with wavelengths less than 91 nm at a rate proportional to their concentration [H] and are de-ionized by exchanging their charges with atomic oxygen, in the reaction

$$H^+ + O \rightarrow H + O^+ \tag{3.2}$$

at a rate proportional to $[H^+]\,[O]$. Thus in the region of photochemical equilibrium

$$[H] \propto [H^+]\,[O] \tag{3.3}$$

If now the scale height of a constituent X is written $H(X)$ expression (3.3) implies that

$$[H^+] \propto \exp\left[-h\{1/H(H) - 1/H(O)\}\right] \tag{3.4}$$

or since $H(O) = (1/16)\,H(H)$

$$[H^+] \propto \exp\{+h/H(O)\} \quad \text{(approximately)} \tag{3.5}$$

representing an upwards *increase* of H^+ ions with a (negative) distribution height roughly equal to the scale height of oxygen. The upwards increase occurs because, the scale height of oxygen being less than that of hydrogen, the rate of loss of H^+ by charge-transfer to O decreases upwards more rapidly than the rate of production.

There is, in addition, another process for the production of H^+ ions that can be more important than photo-ionization; it is the opposite of the reaction (3.2) and involves the transfer of a positive charge from an O^+ ion to a neutral hydrogen atom.† The processes of production and loss are then represented by

$$H + O^+ \rightleftharpoons H^+ + O \tag{3.6}$$

† This reaction is rapid because of a remarkable chance coincidence between the ionization potentials of H and O^+.

and the equilibrium distribution can be calculated by equating the rates at which this reaction proceeds in the two directions. We then have

$$[H^+] \propto [H][O^+][O]^{-1} \qquad (3.7)$$

so that if the different particles are distributed exponentially each with its proper distribution height

$$[H^+] \propto \exp\left[-h\left\{\frac{1}{H(H)} + \frac{1}{H(O^+)} - \frac{1}{H(O)}\right\}\right] \qquad (3.8)$$

or, since $H(O)$, $H(O^+)$ and $H(H)$ are in the ratio $1:2:16$,

$$[H^+] \propto \exp\{+h/2H(O)\} \quad \text{(approximately)} \qquad (3.9)$$

The result is similar to that (3.5) derived on the supposition that the H^+ ions are produced by photo-ionization.

At the greatest heights a photochemical distribution determined by rates of production and loss is replaced by one controlled by diffusion. This diffusion-dominated region has been called the *protonosphere*. In it the electron concentration is described by the expression

$$[e] = [e]_0 \exp\{-h/2H(H)\} \qquad (3.10)$$

Since diffusion occurs freely only along the direction of the magnetic field, $[e]$ and h in (3.10) refer to values, not along a vertical line, but along a field line. The plasma in a given tube of force thus diffuses along it and remains inside it: protons, unlike atoms of neutral hydrogen, do not normally leave the earth's gravitational attraction however rapidly they move. They may, however, travel far from the earth by following distorted geomagnetic lines of force (p. 78).

In the protonosphere there is another type of interaction that is probably important. The positive charge of a proton can fairly easily be passed to a neutral hydrogen atom, so that the movements of charged and neutral hydrogen atoms are not entirely independent. The results of this interaction are complicated and not well understood.

3.2.3 Helium ions [97]
Atmospheric helium is ionized by solar ultraviolet with wavelengths less than 50 nm at a rate proportional to its concentration. The ions

are lost by transferring their charges to neutral atoms (of O) or to molecules (of N_2) by reactions such as

$$He^+ + O \rightarrow O^+ + He \qquad (3.11)$$

which proceeds at a rate proportional to $[He^+][O]$. Under conditions of photochemical equilibrium the rate of production, proportional to $[He]$, is balanced by the rate of loss so that $[He] \propto [He^+][O]$. Hence, if the neutral gases are in diffusive equilibrium

$$[He^+] \propto \exp\left[-h\left\{\frac{1}{H(He)} - \frac{1}{H(O)}\right\}\right]$$

or, since $H(He) = 4H(O)$,

$$[He^+] \propto \exp\{+3h/4H(O)\} \qquad (3.12)$$

Expression (3.12) shows that, under conditions of photochemical equilibrium, the concentration of He^+ ions *increases* upwards just as does the concentration of H^+ ions.

At sufficiently great heights the equilibrium of He^+ ions is determined, not by photochemical processes, but by diffusion. The change-over occurs near the level where the concentration of the ions is diminished equally rapidly by chemical reactions and by diffusion. Above that level the diffusive equilibrium is dominated by electro-static forces, and the ions, accompanied by their electrons, have a distribution height that depends on the concentration of the other ions (H^+ and O^+) (p. 109).

3.3   Energetic photo-electrons. Electron and ion temperatures
[45, 62, 79, 80, 81, 94, 176]

During the process of ionization the energy of a photon goes partly to remove the electron and partly to give it kinetic energy. Conservation of momentum requires that the excess energy is shared between the ion and the electron in the inverse ratio of their masses, so that nearly all goes to the electron. The energy of a photon is about 10 eV at a wavelength of 120 nm and about 40 eV at 30 nm, whereas ionization energies are about 15 eV: the photo-electrons liberated in the F layer thus have kinetic energies of order 10 or 20 eV. They lose this energy by colliding with the atoms and molecules, or with the less energetic ambient electrons that surround them. While the energy of the photo-electron is great it is lost mainly by raising atoms or molecules to higher quantum states through inelastic collisions; but when the energy has been reduced to about 1.5 eV it can be passed to other particles only by means of elastic collisions. If the collision is with a particle of much greater mass, such as a neutral particle or an ion, the fraction of energy transferred is small, whereas if it is with a particle of similar mass, such as an electron, the fraction is large. Most of the remaining energy of the photo-electron is therefore passed to the ambient electrons and their temperature is increased: the increase is greater at times, such as sunrise, when there are fewer electrons to share the energy.

Once the ambient electrons have been heated up they can lose their excess energy by conduction downwards or by heating up the surrounding gas. The rate at which the gas is heated depends on the collision cross-section for momentum transfer: it is much greater for collisions between electrons and ions than for collisions between electrons and neutral particles (it is about 10^5 times greater at 1000 °K, see §6.1). The ambient electrons thus cool mainly by heating the surrounding ions.

At sunrise the temperature of the ambient electrons rises so rapidly that the other processes of energy transfer are too slow to maintain thermal equilibrium between the different types of particle: although the temperature of the electrons increases rapidly, that of the ions and neutral particles remains almost unaltered. A little later the temperature of the ions also increases until it is nearly the same as that of the electrons: still later both return to the temperature of the neutral air which by that time has also increased a little. Changes of this kind can be seen in the results of fig. 3.5.

Sometimes the times of sunrise at two geomagnetically conjugate points differ by one hour or more. Then energetic photo-electrons from the illuminated place can travel, along lines of force, to the place that is in darkness, and can produce an observable increase in the temperature of the ambient electrons there: a second increase is observed when the sun rises locally. An example is shown in fig. 3.6.

The transfer of heat from the ambient electrons to the ions depends

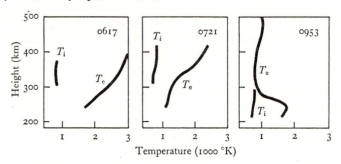

Fig. 3.5. The height-distribution of the temperatures of electrons (T_e) and ions (T_i) at three times in the early morning. Just after sunrise the temperature of the electrons is much increased. Later they cool down by sharing their energy with positive ions, whose temperature is correspondingly increased slightly (after reference 79).

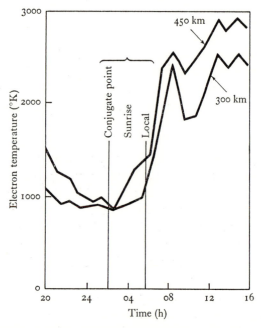

Fig. 3.6. The temperature of electrons in the F region (shown at heights of 300 and 450 km) starts to increase in the morning when the sun first produces photo-electrons at the magnetically conjugate point. There is a further increase when it produces photo-electrons overhead. The two times indicated for sunrise refer to a height of 150 km where photo-electrons are produced most copiously (after reference 62).

on the large cross-section for collisions between them. This cross-section is proportional to the inverse square of the electron temperature. It thus decreases as the electrons are heated up, and if they are heated too rapidly an unstable situation can arise, in which the hotter they are the less they are able to lose their heat. Stability is then re-established only when the temperature of the ambient electrons becomes so great that sufficient heat loss occurs through the other processes of heat conduction and of transfer to neutral particles.

4 The magnetosphere†

4.1 The boundary of the magnetosphere. The magnetopause
[38, 68, 69, 70, 122, 157, 175]

If the solar wind did not exist, the outer ionosphere would consist of a proton-and-electron plasma with concentration determined by diffusion along the magnetic field: it would merge gradually into the plasma of outer space. The situation is completely altered by the presence of the highly conducting solar wind. The currents that are induced in the wind as it moves through the earth's magnetic field give rise to additional fields which add to the geomagnetic field and increase it at the earth's surface as though it had been compressed. Moreover the force exerted by the earth's field on the induced currents causes the wind to change its direction so that it avoids a region surrounding the earth. A sharp boundary then divides an outer region, where the wind continues to flow, from an inner one that contains the compressed geomagnetic field, and from which the wind is excluded. The inner region is called the *magnetosphere* and the boundary the *magnetopause*.

The boundary is roughly spherical on the sunward side and is roughly cylindrical in the anti-sun direction (fig. 4.3). In the direction of the earth–sun line, it is at a distance where the kinetic pressure of the solar wind particles (mass m, velocity V and concentration n) is equal to the magnetic pressure of the (modified) earth's field (B') so that

$$nmV^2 = B'^2/2\mu_0 \qquad (4.1)$$

The situation near the boundary can be understood by considering the motions of the individual particles in the wind. Take an oversimplified model in which protons and electrons, with equal concentrations and velocities, are projected into a uniform magnetic field B perpendicular to their velocity. Suppose that a steady state is set up, as in fig. 4.1, in which a plane PQR, normal to the particles' velocity, divides a region on the right, where there is an enhanced magnetic field $2B$, from a region on the left where the field has been reduced to zero. A current sheet must flow in the direction PQR near this plane

† [4, 19, 22, 25, 75, 76, 142, 155].

to produce a field $-B$ on the left-hand side, to cancel the field that was originally there, and a field $+B$ on the right-hand side to double the original field: the linear current density in this sheet has magnitude $i = 2B/\mu_0$.

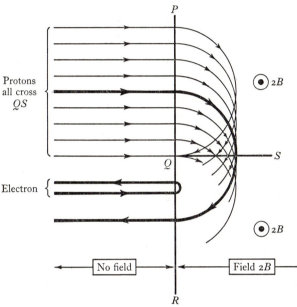

Fig. 4.1. A simple model of the magnetospheric boundary. A uniform magnetic field, B, is directed out of the paper. A stream of electrons and protons moving from the left penetrates to a plane PQR: it sweeps the field out of the space on the left of the plane and 'compresses' it until its strength is doubled in the space on the right. The charged particles moving in semicircles on the right-hand side of PQR constitute a sheet of current, in the direction PR, which produces a magnetic field just sufficient to annul the original field, B, on the left-hand side and to double it on the right-hand side.

After an approaching particle has crossed the boundary and entered the uniform field $2B$ it follows a circular path of radius $r = mV/2eB$ and, after moving round a semicircle, it returns reflected into the stream. Consider a slab of unit thickness perpendicular to the diagram: the protons that cross the area QS in unit time are those that are incident on the area $PQ (= 2r)$ in unit time. (The proton orbits have a radius much greater than the electron orbits so that the protons crossing QS greatly outnumber the electrons.) This flux of protons, $2nVr$ in unit time, represents a current $2nVre$ flowing in a slab of unit thickness perpendicular to the diagram: it is the linear current density j. Insertion of the magnitude of r shows that $j = nmV^2/B$, and, since $j = 2B/\mu_0$ there results

$$2B^2/\mu_0 = nmV^2 \tag{4.2}$$

In the present situation, where the velocity of the particles is reversed at the boundary, and where the (enhanced) magnetic field is $2B$, the pressure-balance type of calculation that was used earlier leads to the expression

$$(2B)^2/2\mu_0 = 2nmV^2 \tag{4.3}$$

Apart from a numerical constant the expressions (4.2) and (4.3) are similar.

Since the protons and electrons penetrate to different distances in the direction QS the resulting electric field should be taken into account and the above discussion is incomplete. It is nevertheless useful for indicating the way in which the current sheet is built up; it also provides an estimate of the sheet's thickness, equal to the radius r of the protons' orbits. By inserting the magnitudes $V = 300\,\mathrm{km\,s^{-1}}$, $B = 10^{-8}\,\mathrm{T}\,(10^{-4}\,\mathrm{gauss})$, and e/m (for protons) $= 10^8\,\mathrm{C\,kg^{-1}}$ into the expression $r = mV/2eB$ this thickness is found to be about $150\,\mathrm{km}$, very small on an astronomical scale.

In the real situation the solar wind impinges, not on a uniform magnetic field, but on the field of the earth's dipole. The essence of the situation along the sun–earth direction can then be discussed (fig. 4.2) by supposing that the sharp boundary is a plane, at a distance d from the centre of the earth, containing a current whose magnetic field on the windward side just cancels the field of the earth's dipole, of moment M (fig. 4.2(a)), and on the earthward side distorts the original field (fig. 4.2(b)). On the windward side the field of the current is thus that of a dipole $-M$ at the centre of the earth, distant d. On the earthward side the field is that of an image dipole of moment $+M$ on the other side of the current sheet and at a distance d from it or $2d$ from the centre of the earth.

Let the total fields at the surface of the earth (radius R) and at the current sheet (distance d) be represented by B_R and B_d, and let B_{R_0} and B_{d_0} represent the fields at these distances due to the earth's dipole

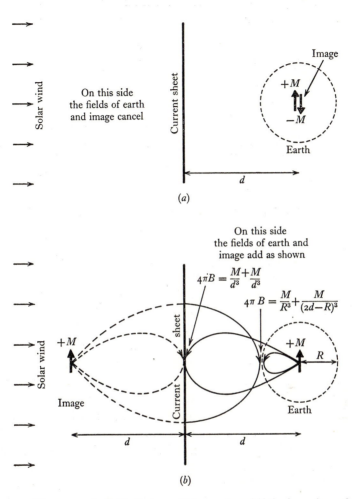

Fig. 4.2. The magnetic fields that would be produced if the boundary of the magnetosphere were plane. The earth's field is that of a dipole of moment $+M$ at its centre, currents flow in the plane boundary to produce a field that on the sunward side seems to come from a dipole $-M$ at the earth's centre, and on the earthward side to come from an image dipole $+M$ at the position of the earth's image in the plane of the current. The resulting total fields are indicated.

alone. Then $B_{d_0} = B_{R_0}(R/d)^3$ and this field is doubled by the presence of the current sheet so that

$$B_d = 2B_{d_0} = 2B_{R_0}(R/d)^3 \qquad (4.4)$$

At the earth's surface the field $B_{R_0}(= M/4\pi R^3)$ of the earth alone is increased by the field $\Delta B_R = M/4\pi(2d-R)^3$ of the image dipole. Thus

$$\Delta B_R = B_{R_0}\{R/(2d-R)\}^3 = B_{R_0}(R/2d)^3 \qquad (4.5)$$

when the field of the image dipole is added.

The distance (d) of the current sheet can be found by inserting the parameters of the solar wind into (4.1) to give B' and putting $B' \equiv B_d$ into (4.4). Then with

$$n = 5 \times 10^6 \, \text{m}^{-3}, \quad V = 3 \times 10^5 \, \text{m s}^{-1}, \quad m = 1.7 \times 10^{-27} \, \text{kg}$$

and $$B_{R_0} = 3 \times 10^{-5} \, \text{T}$$

it is found that $d = 10R$. Insertion of this value into (4.5) then shows that $\Delta B_R = 3 \times 10^{-9} \, \text{T}$ (3 gamma). This calculation thus leads to the conclusion that the current sheet is at a distance of about 10 earth radii and that the geomagnetic field is compressed so that its magnitude at the earth's surface is increased by about $3 \times 10^{-9} \, \text{T}$.

Simple considerations like these show the kind of behaviour expected in the sun–earth direction. Similar considerations can be applied in more detail to investigate the three-dimensional shape of the magnetospheric boundary. It is then found that the compression of the field is somewhat greater, but the distance of the boundary in the sunward direction is not much altered. It is usual to suppose that in addition to particle pressure, normal to the boundary, the solar wind exerts a tangential force on the boundary on the side away from the sun, and drags it out into a long tail, the *magnetotail*, in the anti-solar direction. The magnetic field lines, embedded in the plasma, then take the form shown in fig. 4.3.

In this field there are neutral points near A and B, and a neutral sheet at CD, where the field directions reverse in comparatively short distances. It has been suggested that neutral points, like A and B, could possibly provide openings through which charged particles could enter the magnetosphere from outside. Thus if a particle is moving along a field line in outer space, in a suitable cyclotron orbit large enough to intersect a different field line belonging to the magneto-sphere, it could attach itself to that line and become trapped on it.

The neutral sheet *CD* is probably important for a different reason. When neighbouring lines of force run in opposite directions, there is a possibility that they may change their shapes in the way suggested in fig. 4.4: a change of that kind is called a reconnection of the lines. It arises from a change in the distribution of current, and is accompanied by a flow of the plasma, as indicated by the thick lines in

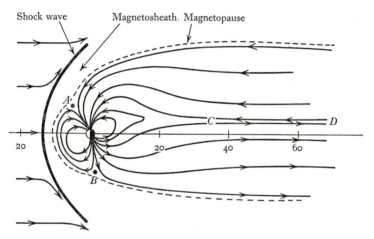

Fig. 4.3. The geomagnetic field lines distorted by the action of the solar wind. The dotted line indicates the magnetopause, inside which the magnetic field is confined. There are neutral points at *A* and *B* and a neutral sheet at *CD*. Geocentric distances are indicated, in units of earth radii, along the sun–earth line (after reference 19).

fig. 4.4 (*b*). The precise mechanism is obscure, but it has been suggested that it may result in the acceleration of particles to considerable energies and their subsequent flow along field lines into the lower ionosphere. In this process the energy comes from the current which ceases when the neutral sheet disappears.

Evidence is accumulating to show that the behaviour of particles in the magnetosphere depends on the direction of the component of the interplanetary magnetic field that is parallel to the earth's dipole axis. Fig. 4.5 shows at (*a*) how a field component parallel to the dipole axis combines, in free space, with the dipole field to produce a neutral circle in the equatorial plane of the dipole when the component field is directed to the south, and at (*b*) how it forms two neutral points above the two poles when it is directed towards the north.† The situation at (*a*) where the field component is directed southwards is the interesting one.

When the interplanetary field is transported with the solar wind the lines of force shown at (*a*) are distorted to take the shapes shown at (*c*): because they move with the wind, lines o that pass through the neutral

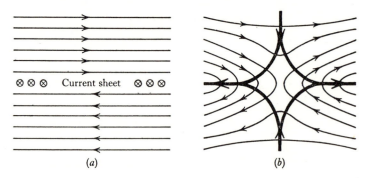

(*a*)   (*b*)

Fig. 4.4. To illustrate field line reconnection. The field represented by the lines in (*a*), accompanied by a current sheet as shown, can be converted into that represented by the thin lines in (*b*). The change is accompanied by movements of plasma, as indicated by the thick lines, and by acceleration of the charges.

point *P* take, at subsequent times, the forms shown at 1, 2, 3, ... The terrestrial and the interplanetary field lines 'merge' at *P* and the resulting line is swept back towards the magnetotail until lines from the two sides merge again at *Q*. Meanwhile lines from the earth move out to *P*, to replace those that have merged, and at *Q* they move inwards, towards the earth; they carry the plasma with them to produce a resultant flow indicated by the arrows. There is no corresponding situation when the interplanetary field component is directed northwards.

Dungey [74] first drew attention to the merging of field lines in this way, and the consequent transfer of plasma from the magnetotail to the

† The direction of the interplanetary field in a plane perpendicular to the dipole axes was shown in fig. 1.5.

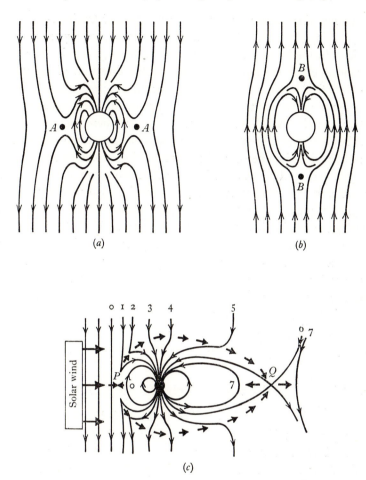

(a)

(b)

(c)

Fig. 4.5. The interaction of the earth's dipole magnetic field with the N–S component of the interplanetary field. (*a*) In free space a uniform field directed as shown (i.e. towards the *south* pole of the earth) combines with the dipole field to produce a neutral ring that intersects the plane of the diagram at *AA*. (*b*) If it has the opposite direction it forms two neutral points *B* and *B* above the two poles. If the field shown in (*a*) is carried from left to right by the solar wind the field lines are distorted so that they take, in succession, the positions labelled o, 1, 2, 3 ... in (*c*). Plasma is carried with them as indicated by the small dark arrows and is injected into the magnetosphere at the neutral point *Q* to travel towards the earth's polar regions. (*c* is after reference 21.)

earth when there was a southwards component of the interplanetary field, and suggested that it might account for the depositing of energetic particles in the ionosphere. Recent observations [169], showing that substorms often accompany a change in the direction of the interplanetary field from north to south, support this suggestion (see §5.2.2).

When the solar wind impinges on the magnetosphere it behaves in many ways like a wind of neutral air striking a spherical object. If the speed of the wind is greater than the speed of waves that can travel in it the behaviour is supersonic. In the solar wind two kinds of wave are possible, one a sound wave with a velocity of about $5 \times 10^4 \, \mathrm{m \, s^{-1}}$, and the other a hydromagnetic (Alfvén) wave (p. 153) with a velocity $B(\mu_0 n m_i)^{-\frac{1}{2}}$; with $B = 10^{-9} \, \mathrm{T}$ this velocity is about $10^4 \, \mathrm{m \, s^{-1}}$. Since the velocity of the wind is greater than either, its motion is supersonic.

The impact of the solar wind on the magnetospheric boundary thus results in the setting up of a stand-off shock wave near which the flow lines change their direction abruptly. In this wave, electromagnetic forces between the particles play a part similar to collisions in the more usual kind of shock wave in a neutral gas. Between the shock wave front and the magnetopause there is a region in which the particles have an irregular motion: it is called the *magnetosheath* (see fig. 4.3).

4.2   Charged particles in the highest ionosphere [36, 135]

4.2.1   The plasmasphere [63, 64, 65, 74, 114, 131, 152, 161]

Throughout the ionosphere the majority of charged particles are produced by the ionizing action of solar radiation, they have energies of about 0.1 eV corresponding to a temperature of a few thousand degrees kelvin. Their concentration varies with height above the geomagnetic equator as shown in fig. 4.6: there is an inner region, extending to about 4 earth radii, where the concentration is of order $10^9 \, \mathrm{m^{-3}}$ and an outer one where it is smaller, of order $10^7 \, \mathrm{m^{-3}}$. The sharp radial gradient in the concentration has sometimes been called a 'knee'.

Since the ionospheric plasma is in diffusive equilibrium along geomagnetic field lines there is a sharp gradient in the electron concentration in other parts of the ionosphere that lie on the field line coincident with the knee: there is thus an inner region of greater concentration, separated from an outer region of lesser concentration (see fig. 9.13, p. 206). The inner region is called the *plasmasphere*, the boundary between it and the outer magnetosphere being called *the plasmapause*. When it is desirable to define the inner boundary of the magnetosphere it is convenient to identify it with the plasmapause.

The intersection of the plasmapause with the F layer has been located, at night, with the help of topside sounders. Fig. 4.7 shows the

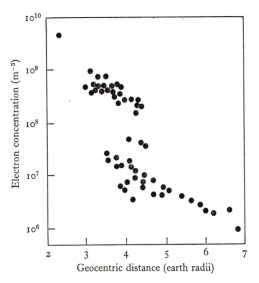

Fig. 4.6. The electron concentration as a function of geocentric distance above the equator as deduced from observations of whistlers (p. 206). The sharp gradient near four earth radii represents the plasmapause (after reference 63).

distribution of electrons in the upper part of the layer as a function of latitude: the marked gradient of concentration near latitude 50° N corresponds, when projected back along a line of force, to a plasmapause at an equatorial distance of 4 earth radii. This sharp latitudinal minimum of electron concentration has been called the *mid-latitude trough*.

The difference between the electron concentrations on the two sides of the plasmapause has been explained in terms of plasma movements caused by electric fields and by diffusion. The fields that are important originate in two different places, some in the low ionosphere, and some in the boundary of the magnetosphere: they are transferred, along the highly conducting geomagnetic field lines, to other places, where they result in movements of the plasma. For reasons to be explained shortly these movements are different on the two sides of the plasmapause; they lead to a loss of plasma from the outer part of the ionosphere but not from inside the plasmasphere.

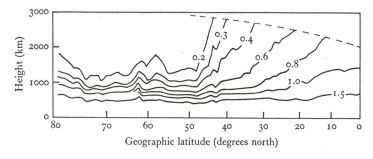

Fig. 4.7. The plasma frequency f_0 (in MHz) at different heights and latitudes measured with the help of a topside sounder. The electron concentration per cubic metre is equal to $1.24 \times 10^{10} f_0^2$. The sharp decrease of concentration near latitude $50°$ marks the position of the plasmapause at the relevant heights: a field line leaving the earth near this place passes over the geomagnetic equator at a geocentric distance of 4 earth radii.

Electric fields originating near the boundary of the magnetosphere, transferred along field lines to other parts of the magnetosphere, produce convective movements of the kind shown in fig. 4.8(a). The field of the atmospheric dynamo, in the low ionosphere, transferred to the rest of the magnetosphere, causes it to rotate with the earth. This rotation can equally well be described by saying that the geomagnetic field lines are carried round by the conducting ionosphere at their base, and carry the plasma with them. When this rotation of the plasma near the earth is added to the convective motion of fig. 4.8(a) the resulting motion in the equatorial plane of the magnetosphere is as shown in fig. 4.8(b).

The field lines, carried with the moving plasma of fig. 4.8(b), move in three dimensions, as shown in fig. 4.9. Those in the inner group, rotate with, and remain comparatively near, the earth, but those in the outer group, rotating with the convection of the magnetospheric plasma, are carried away to great distances in the tail.

The movements of the plasma illustrated in figs. 4.8 and 4.9 are caused by electric fields and are therefore *perpendicular* to the field

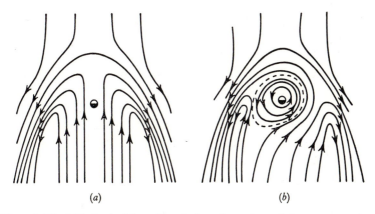

Fig. 4.8. The 'friction' of the solar wind on the surface of the magnetosphere would produce the plasma movements indicated at (*a*) if the earth were stationary. But the rotating earth carries the nearby plasma with it so that the resulting movements are as indicated at (*b*). Inside the dotted curve the plasma remains near the earth, whereas outside it is carried far away: the plasmasphere is inside the curve (after reference 130).

Fig. 4.9. Magnetic field lines from the polar regions are carried far out into the magnetotail by the convection of the magnetospheric plasma: those from lower latitudes remain nearer the earth as they rotate with it.

lines; there is, in addition, a drift *along* the field lines, caused by diffusion. The plasma supplied by the ionization of the air at comparatively small heights diffuses upwards along the moving field lines: in the inner region it remains on closed lines, carried round by the earth's rotation, and is undiminished in content; in the outer region it is tied to lines of force that are carried out into the magnetotail: it diffuses along them to great distances where some of it escapes into outer space.

The plasma that can escape in this way is on field lines that terminate in high latitudes: the resulting movement of plasma away from these regions has been called a *polar wind*. It results in depletion of the ionization in a region whose boundary, in the equatorial plane, is represented by the dotted line in fig. 4.8(b): in the region inside this boundary there is no corresponding loss of ionization. It has been suggested that this boundary forms the plasmapause that divides the inner region where the plasma is more concentrated from the outer region where it is less concentrated. In the equatorial plane, shown in fig. 4.8(b) it is at the distance, about 4 earth radii, of the sharp gradient shown in fig. 4.6.

4.2.2   Trapped particles [19, 23, 36, 135]

In some parts of the magnetosphere the ions and electrons that have energies of order 0.1 eV, corresponding to temperatures of a few thousand degrees, are accompanied by others with much larger energies, of order several keV to several MeV. The particles with smaller energy have concentrations of order $10^9\,\mathrm{m}^{-3}$ in the plasmasphere and $10^7\,\mathrm{m}^{-3}$ outside, so that they represent energy densities of about $10^8\,\mathrm{eV\,m}^{-3}$ and $10^6\,\mathrm{eV\,m}^{-3}$. Although those with larger energy have concentrations only of order $10^4\,\mathrm{m}^{-3}$ they correspond to energy densities of order 10^7 to at least $10^{10}\,\mathrm{eV\,m}^{-3}$ and make an important contribution to the energy of the magnetosphere.

Charged particles of both kinds collide with the surrounding magnetospheric protons with collision cross-sections (σ) depending on their energies (expressed in terms of equivalent temperature T, where 0.1 eV corresponds to $T = 1160\,°\mathrm{K}$) roughly as given by the expression (see (6.9)) $\sigma = 6 \times 10^{-9} \times T^{-2}\,\mathrm{m}^2$ so that for a particle of energy 0.1 eV (1000 degK) it is about $6 \times 10^{-15}\,\mathrm{m}^2$ and for a particle of energy 1 keV about $6 \times 10^{-23}\,\mathrm{m}^2$: it is even smaller for a particle with still more energy. A charged particle moving amongst the magnetospheric protons, with a concentration of order $10^7\,\mathrm{m}^{-3}$, thus travels enormous distances (of order 10^4 km and 10^{12} km for energies of 0.1 eV and 1 keV) before it makes a collision. During these very long paths of free travel it can become trapped so that it moves in a helical path around a geomagnetic field line and bounces back and forth between two *mirror points* one at the north, and one at the south, end of its path (§7.3). The period of gyration about a field line is of order 10^{-6} s for electrons and 10^{-3} s for ions. The period of bouncing back and fourth is the same for electrons and protons; on a field line that crosses the equator at a height of 2000 km it is about 0.5 s. The helical motion as a whole moves slowly sideways round the earth towards the east for electrons and towards the west for protons. The time for a complete traverse of the earth depends on the height and on the energy of the particle. For the height previously considered it is about 10 h for particles of energy 50 keV, and 3 min for 10 MeV. The oppositely directed drifts of the proton and electron helices constitute a *ring current* round the earth, flowing towards the west. The magnetic field that this current produces near the surface of the earth is discussed in §7.3.

As a trapped particle continues to move, bouncing back and forth on a helical path that itself gradually moves round the earth, it covers, in the course of time, a surface or 'shell'. It is usual to denote the size and shape of one of these shells by the magnitude of a quantity L that defines its average geocentric distance at the geomagnetic equator, measured in terms of the earth's radius as a unit. If the magnetic field of the earth were symmetrical a given L-shell would be symmetrical also, but because there are departures from symmetry a trapped particle moves on a distorted surface whose shape is determined by the actual geomagnetic field. A given L-shell lies at a distance above the equator where the field is constant, and elsewhere its shape is such that the field lines lie in it. Even when the surfaces are distorted they are labelled by the magnitude of L at the geomagnetic equator.

The depth to which a particle descends at its mirror point depends on the closeness of the turns in its helical path as it crosses the equator, where the geomagnetic field is horizontal. This closeness is measured in terms of the pitch angle (α) between the magnetic field and the particle's velocity. A particle crossing the equator (where the magnetic field is B_0) with pitch angle α follows a field line down towards the earth until the field strength reaches a magnitude B_m such that

$\sin^2 \alpha = B_0/B_m$: the motion along the field line then ceases momentarily, the mirror point has been reached, and the particle returns along its path. It should be noted that the depth of the mirror point does not depend on the nature of the particle or on its energy but only on the pitch angle of its motion and the shape of the magnetic field. The field at the surface of the earth has anomalously small values in the South Atlantic ocean, so that there, the mirror points are much lower than elsewhere.

A particle continues to move over a magnetic shell so long as it experiences only forces arising from the steady magnetic field. Its motion can be disturbed by collisions with other particles, by hydromagnetic waves, or by changes in the geomagnetic field. Collisions act by altering the particle's direction of travel (scattering). Because of the difference in mass an electron is more easily scattered than a proton. As an electron penetrates more deeply into the atmosphere the chance increases that scattering may change its pitch angle so that its mirror point is lower down in the denser atmosphere where it can impart its energy to other particles. This kind of scattering is responsible for a continuous removal of those trapped electrons that have small pitch angles. It occurs most readily at places where the mirror points are lowest: electrons are thus more copiously deposited from the trapping regions in the neighbourhood of the South Atlantic geomagnetic anomaly.

Protons, which are less easily scattered, are removed from the trapping zones mainly by interacting with other particles. In one such interaction they pass their charges to neutral hydrogen atoms by the reaction

$$H^+ \text{(energetic)} + H \rightarrow H \text{(energetic)} + H^+$$

The result is two-fold, first there is a reduction in the number of energetic protons, and second the direction of the new proton is not the same as that of the original one so that its mirror point may be lower, where there are even more interactions. Protons with energy greater than about 400 keV can also be removed by producing nuclear reactions in the atoms they encounter.

It is a curious fact that in the outermost parts of the ionosphere there are many trapped electrons, but comparatively few protons, with great energy. Several processes that might remove protons but not electrons have been suggested in explanation. One of the most likely invokes the interaction between a proton, moving in a helix along a line of force, and the rotating magnetic field in a circularly polarized hydromagnetic wave travelling along the same line of force. If the speed of the proton along the direction of the line is such that the Doppler-shifted wave frequency is equal to the frequency of gyration of the proton, then under some circumstances the proton can lose energy to the wave (§ 8.7.1).

Particles of all energies can be trapped on field lines provided they make few enough collisions. Amongst the particles with energies of 1 eV or less in the background ionosphere, there is little to distinguish the trapped from the non-trapped. The much less numerous particles of greater energy, 1 keV or more, exist only as trapped particles, for when they are removed from the trap into the lower ionosphere they lose their energy and join the others. For this reason trapping has usually been discussed with relation to energetic particles; indeed, it is usual to refer to trapped radiation belts (or *Van Allen belts* after their discoverer) that contain protons and electrons of comparatively great energy.

The high-energy trapped particles originate in several different processes: some are produced by the acceleration of low-energy particles already present in the magnetosphere, some are introduced by naturally occurring nuclear reactions, and some have been introduced artificially by the explosion of atomic bombs at high altitudes.

One important source of protons and electrons is illustrated in fig. 4.10. A (galactic) cosmic ray particle penetrating deeply into the atmosphere causes a nuclear disintegration and liberates a neutron which then travels into the magnetosphere and decays to produce an electron and a proton. The disappearance of the neutron's mass provides energy of about 300 keV to the resulting particles: each also starts life with a velocity (V) roughly equal to that of the neutron, in virtue of which it has energy $\frac{1}{2}mV^2$. Because of the small mass of the electron this extra energy is negligible compared with the disintegration energy of order 300 keV. But for a proton the kinetic energy is more important than the disintegration energy, so that its final energy may be of order 10 MeV.

It has been suggested [1] that energetic protons may sometimes be produced, during solar disturbances, by a stream of neutral hydrogen emitted from the sun with great speed. It is supposed that, as it passes through the magnetosphere, a rapidly moving hydrogen atom sometimes picks up a charge from a (low energy) magnetospheric proton, according to the reaction

$$H\,(rapid) + H^+ \rightarrow H^+\,(rapid) + H$$

and is transformed into an energetic proton.

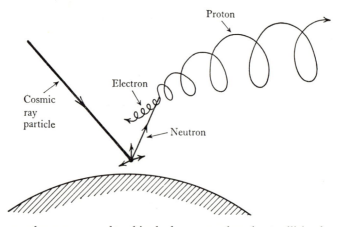

Fig. 4.10. A neutron, produced in the low atmosphere by a collision between a cosmic ray particle and a nucleus, can travel into the upper atmosphere before it decays to form an electron–proton pair. Most of the neutron's kinetic energy goes to produce a high-energy proton which can become trapped in the geomagnetic field.

Protons with energy of a few hundred MeV are sometimes emitted from solar disturbances (solar proton events). They are guided to the polar regions by the geomagnetic field, and there, just like galactic cosmic ray particles, they can produce neutrons in the lower atmosphere that in turn produce energetic protons and electrons by the processes indicated in fig. 4.10. The resulting protons differ from those originating in galactic cosmic rays first by having smaller energies and second by being concentrated near the polar regions.

5 Geomagnetism, ionospheric currents and ionospheric storms†

5.1 Currents in the ionosphere

The magnetic field of the earth approximates to that of a dipole at its centre with strength and direction changing slowly. The slow (secular) changes are accompanied by more rapid changes produced by currents in the ionosphere at small or great (magnetospheric) heights, or by hydromagnetic waves. These currents, in turn, are produced by mechanical forces or electric fields, and their magnitudes depend on the conductivity of the ionized medium in which they flow. They can be divided into:

(i) currents near a height of 110 km; driven by movement of the neutral air: the *atmospheric dynamo*,

(ii) Currents near a height of 110 km, in high latitude regions, driven by forces originating high in the ionosphere: the *polar current system*,

(iii) currents in the ionosphere (magnetosphere) at a geocentric distance of about 4 earth radii: the *ring current*,

(iv) currents in the magnetopause: resulting in a compression of the geomagnetic field.

The magnitude of the ring current, or of the compression of the field, changes only if the energy of the trapped particles, or the strength of the solar wind, changes: the associated magnetic fields do not change much except at times of storms. Even in quiet, non-storm, times, however, these constant fields may be asymmetrically distributed so that to an observer rotating with the earth they may appear to vary during 24 hours [136]. Fields of that kind, that would appear constant to a stationary observer are not discussed any further at present: they are considered in more detail when storms are discussed in § 5.2.2.

† [5, 22, 24].

5.1.1 The atmospheric dynamo and motor [115, 117, 119, 145, 156, 160]

The conductivity of the ionosphere, depending on the motions of the charges in it, is fundamentally different according as the collision frequencies (ν) of the electrons and ions are greater or less than their angular gyro-frequencies (Ω). As height decreases through the middle ionosphere, from 300 to 70 km, the collision frequency of electrons increases from about $10^3 \, \text{s}^{-1}$ to about $10^7 \, \text{s}^{-1}$ and of ions from about $0.5 \, \text{s}^{-1}$ to about $10^5 \, \text{s}^{-1}$, whereas the angular gyro-frequencies remain approximately constant, about $8 \times 10^6 \, \text{s}^{-1}$ for electrons and $160 \, \text{s}^{-1}$ for ions.† Within this height range, therefore, the behaviour of each type of particle changes radically, with the result that a layer of relatively good direct-current conductivity exists at heights around 110 km (§ 7.2.2). The next few paragraphs show how a conducting layer of this kind can behave as the armature of an *atmospheric dynamo* to produce the observed regular changes in geomagnetism that are associated with the solar and the lunar day.

The conducting layer is moved across the earth's magnetic field by the influences of the sun and the moon. The sun acts by heating the ozone layer, whereas the moon acts through its gravitational attraction. The heat input from the sun has a period of one solar day, and a strong harmonic component with period equal to half a day. The distribution of the atmosphere is such that over most of the earth it responds to this component to produce horizontal air movements with a period of 12 hours; they are called a solar semi-diurnal tide. A tide is also driven by the gravitational attraction of the moon; like the tides in the oceans, it has a period of half a lunar day. Fig. 5.1(*b*) shows the horizontal air movements in the lunar tide, and how they are related to the corresponding pressure changes shown in fig. 5.1(*a*).

If, like the lunar tide, the atmospheric solar tide were the result of gravitational attraction it might be expected to be smaller, because of the smaller gravitational force: the situation would then be as it is in the oceans where the lunar tide is the more important. In the atmosphere, however, the thermal origin of the solar tide, coupled with a

---

† At greater heights the height-variation of the geomagnetic field results in important changes in the gyro-frequencies.

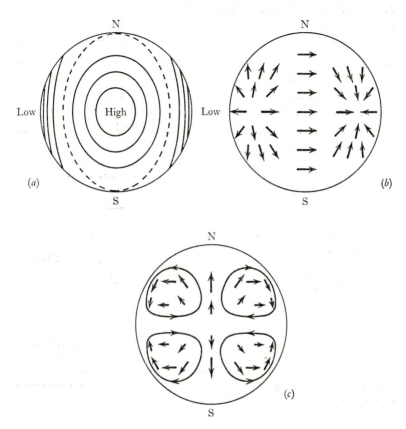

Fig. 5.1. The lunar tide in the atmospheric dynamo. The diagrams represent the earth as seen from the moon, supposed to be in the plane of the equator. (a) The isobars show the distribution of pressure between regions of high and low pressure; (b) as the pattern of pressure moves, with the moon, from east to west, the air in the E region moves horizontally with velocities indicated by the arrows. The nature of the atmospheric oscillations is such that these velocities are opposite to those at the ground; (c) the horizontal air velocities of (b), combined with the earth's dipole magnetic field, induce electric fields indicated by the heavy arrows. They are perpendicular to the wind and have a magnitude proportional to the velocity and to the vertical component of the field. The electric fields produce circulating currents indicated by the closed loops. The magnetic field of these currents constitutes the L_q semi-diurnal variation of the geomagnetic field (after reference 5).

quasi-resonance in the atmosphere, cause that tide to be greater than the lunar one.

Movement of the conducting layer across the permanent magnetic field of the earth induces e.m.f.s in it whose components, in the plane of the layer, have directions perpendicular to the horizontal component of the tidal velocity and magnitudes proportional to the product of that component and the vertical component of the geomagnetic field. Fig. 5.1(c) shows these horizontal e.m.f.s induced by the lunar tide

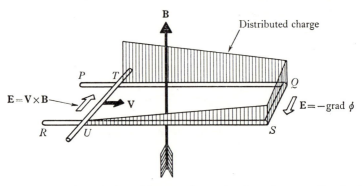

Fig. 5.2. A simple analogy to illustrate the mechanism of current flow in the atmospheric dynamo. When the wire TU moves with velocity $\mathbf{V}$ an electric field $\mathbf{V} \times \mathbf{B}$ is induced in it and current flows round the loop $TQSU$. In the stationary parts TQ, QS, and SU the current is driven by the electrostatic field of a charge distributed along the wires.

and the currents that they produce. The magnetic fields of these currents produce the lunar daily fluctuations of geomagnetism observed at the surface of the earth. During magnetically quiet times the fluctuations assume a fairly regular form; they are called lunar quiet (L_q) variations; similar currents produced by the solar tide are responsible for the solar quiet (S_q) variations.

It is noticeable, in fig. 5.1(c), that the currents do not everywhere flow along the directions of the induced e.m.f.s: it is interesting to discuss the situation in more detail. The simple analogy of fig. 5.2 helps the explanation. Here PQ and RS represent parallel wires connected by a conductor at the right-hand end and supporting at the left-hand end a wire TU that is free to move. There is a magnetic field $\mathbf{B}$ perpendicular to the plane *PQSR*, so that when *TU* is moved perpendicular to its own length with velocity **V** an e.m.f., **V** × **B**, is induced in it and currents flow round the loop *TQSU*. The magnetic field of this current adds to, and modifies, the original field **B** and the situation is analogous to that in the atmospheric dynamo. The question now arises 'What causes the flow of electrons in the different parts of the loop *TQSU*?' In the moving portion *TU* it is the electrodynamic field **V** × **B**: in the stationary part it is an electrostatic field produced by a distribution of space charge that is set up along the wire. If ϕ is the potential set up by the space charge, the field driving the electron in the stationary part of the circuit is the electrostatic field **E** = – grad ϕ; in the moving part it is **V** × **B**.

In this simple example the induced field (**V** × **B**) is localized in one part of the circuit and the electrostatic field (– grad ϕ) in another. In the atmospheric dynamo, induced fields exist all round the current loops, but their magnitudes (and directions) vary. A distributed space charge is then set up that provides an electric potential ϕ, and the total field, **E** = – grad ϕ + **V** × **B**, drives a current whose magnitude and direction depend on the (anisotropic) conductivity of the ionosphere. At places near the equator **V** × **B** is small because **B** is nearly horizontal, and the current is driven almost entirely by the field of the space charge.

In the simple model described above it was supposed that the conducting layer had the form of an isolated slab bounded above and below by horizontal planes, and that only the horizontal components of the fields were responsible for driving the horizontal currents. In a model of that kind vertical components of induced fields (arising from horizontal movement across the horizontal component of the earth's field) produce space charges on the upper and lower boundaries and the resulting electrostatic potential must be included in ϕ. It is, however, usual to discuss the effect of these space charges by considering the way in which they modify the tensor conductivity of the ionospheric slab. The situation is discussed in detail in § 7.2.2 where it is shown that the effective conductivity in the horizontal plane becomes comparatively large near the geomagnetic equator, where the field is horizontal. The current in the atmospheric dynamo is correspondingly great there, it has been called the *equatorial electrojet*.

4   RIT

It is also shown in §7.2.2 that the conductivity, at the equator and elsewhere, has a peak magnitude at a height near 110 km: experiments with rockets have confirmed that the currents of the dynamo flow near that height.

The over-simple model of the ionosphere in which the atmospheric dynamo is supposed to be located in a layer near 110 km, bounded above and below, leads, in the following way, to the idea of an *atmospheric motor* at greater heights, in the F region, driven by the dynamo below. It is supposed that the conductivity of that part of the ionosphere above the dynamo is much greater along the direction of the magnetic field than across it, so that electrostatic fields developed in the dynamo region are transferred along sloping lines of force to the F region at heights around 250 km. These fields, acting in conjunction with the geomagnetic field, can produce movements of the F region. Since the collision frequencies of ions and electrons are much less than the corresponding gyro-frequencies at these heights the electric field (**E**) moves the ions and electrons together, in a direction perpendicular to the magnetic field (**B**), with a velocity $\mathbf{V} = \mathbf{E} \times \mathbf{B}/|\mathbf{B}|^2$ (p. 122). This movement has important consequences near the equator (§ 3.1.2).

A detailed theory of the atmospheric dynamo would be very complicated; it would involve a knowledge of the world-wide distribution of ionization, calculation of the height-variation of the tensor conductivity over the earth, a determination of the solar and lunar tidal variations at all points on the earth, and a calculation of the currents that flow both horizontally and vertically. Only partial success has so far been achieved, mainly by theories based on simplifying assumptions of the kind outlined here.

5.1.2   The polar current system [38, 42, 55, 59, 104, 117]

The solar daily variation of magnetism observed at middle and low latitudes in magnetically quiet times (the S_q variation) can be explained in terms of the atmospheric dynamo. In polar regions there is an additional solar variation, denoted by S_q^p, that can be attributed to the current system shown in fig. 5.3 flowing at heights near 112 km. Unlike the current in the atmospheric dynamo, it is driven by electric fields arising in the magnetosphere, and is most simply understood in terms of the concept of frozen-in lines of force (p. 131) as follows. When plasma moves far out in the magnetosphere it causes the lines of magnetic force to move with it, at all levels where collisions are un-important. Fig. 5.4 shows how a movement far out in the magneto-sphere is transferred along the moving field lines, to the low ionosphere at high latitudes.

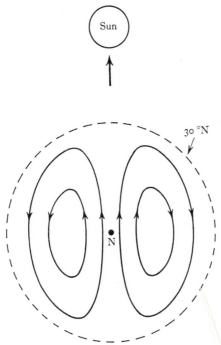

Fig. 5.3. An idealized representation of the current system in the polar region that is responsible for the S_q^p part of the solar diurnal variation.

The charged particles follow the moving lines of force so long as the collision frequency is less than the gyro-frequency. As a line penetrates further into the ionosphere this condition ceases to be valid first for ions at a height of about 140 km, and then, lower down at about 80 km, for electrons. There is thus a region between heights of 140 and 80 km where electrons are dragged round by the lines of force whereas ions are not. The currents, flowing in a direction opposite to the electrons' motions, constitute the polar current system of fig. 5.3.

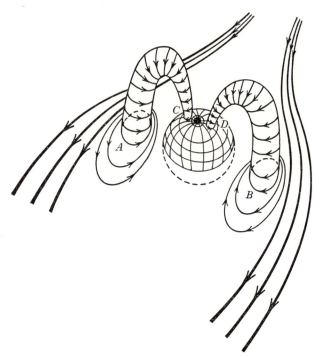

Fig. 5.4. To show how the current system of fig. 5.3 is driven by motions of plasma in the distant magnetosphere. The solar wind produces circulatory movements of plasma at A and B. These carry the magnetic field lines on the tubes AC and BD with them. When these field lines enter the low ionosphere at C and D where $\nu_e \ll \Omega_e$ but $\nu_i \gg \Omega_i$ they carry the electrons, but not the ions, with them. The circulating loops of electrons at C and D correspond to currents in the opposite directions as shown in fig. 5.3.

5.2   Solar disturbances: ionospheric and magnetic storms
[6, 41, 68, 70, 88, 120, 133, 134]

Sometimes *solar disturbances* occur accompanied by enhancement of the solar photon radiation in parts of the spectrum; or by an increase in the velocity and concentration of the solar wind to such an extent that the modified portion behaves like a cloud of denser plasma; or by the ejection of more energetic protons and electrons with such small concentrations that they behave not as plasma but as independent charged particles. The three types of disturbance sometimes occur together and sometimes separately. When the photons, or the particles, reach the vicinity of the earth they produce phenomena known variously as sudden ionospheric disturbances, ionospheric storms, or magnetic storms.

5.2.1 Sudden ionospheric disturbances (SIDs)

The most easily observable evidence of a solar disturbance is enhancement of the visible H-alpha line over a small region usually near a complex group of sunspots: it is called a *solar flare*. When it occurs the solar X-radiation with wavelength less than about 1 nm increases in strength so as to produce increased ionization in the D region at heights around 80 km (fig. 2.3). The increase in electron concentration leads to several phenomena grouped together under the name *sudden ionospheric disturbance (SID)*. The disturbance has a profound effect on the travel of radio waves, and a less important one on geomagnetism. Because SIDs are produced by photon radiation from the sun they occur only on the sunlit side of the earth and their effects are most intense when the sun is near the zenith.

It is interesting to note that, although the H-alpha line and the short wavelength X-radiation are considerably enhanced during a solar flare, the strength of the Lyman-alpha line in the ultraviolet usually changes very little. On some occasions increases in the electron concentrations of the E and F layers have been observed during solar flares, presumably because the corresponding part of the ultraviolet spectrum has then increased, but there have not yet been enough measurements in satellites to establish clearly whether that increase is common.

During an SID the electron content of the D region increases very suddenly within a few minutes and afterwards slowly recovers to its normal magnitude within a time of order $\frac{3}{4}$ to $1\frac{1}{2}$ hours. Attempts have been made to estimate the time constant with which the electron concentration decays by observing the rate at which some quantity such as radiowave absorption returns to normal after the disturbance. They have not been very successful, however, because assumptions have to be made about the height at which the electrons are situated and about the way in which the intensity of the ionizing radiation changes.

During an SID the absorption of medium and high frequency radio waves is often great enough to result in a complete removal of a signal reflected from the E and F layers, it is then said that there is a *short wave fadeout* (SWF).

The intensities of low and very low frequency waves change during an SID in ways that depend on the frequency and on the angle of reflection. For example waves of frequency near 27 kHz, reflected obliquely from the ionosphere, are increased in strength; the increase is clearly noticeable when atmospherics, originating in distant lightning flashes, are observed. These *sudden enhancements of atmospherics* (SEAs) have proved useful at observatories where it is desired to maintain a warning system for the occurrence of visible solar flares. On waves of very low frequency reflected from the ionosphere the most clearly marked effect of an SID is a sudden change of phase, usually called a *sudden phase anomaly* (SPA).

When an intense SID occurs it is frequently found that there is a simultaneous small change, sometimes called a crochet, in the geomagnetic field. The theory of the atmospheric dynamo would account for a change of that kind if the conductivity were increased at the level of the dynamo currents, indeed it has been noticed that the changes during a crochet correspond, at several places, to an increase of the normal daily solar variation (S_q) of the field. Explanation of the crochet in terms of the dynamo is, however, not simple because the increase in ionization during an SID is thought to be mainly at a height near 90 km whereas the currents of the dynamo are thought to flow at heights near 110 km.

5.2.2 Ionospheric and magnetic storms (polar regions)

[1, 40, 42, 70, 105, 117, 133, 134, 140]

A solar disturbance is frequently characterized by an enhancement of the solar wind: the concentration of particles is increased to about $10^7 \, \mathrm{m^{-3}}$ (instead of 5×10^6 in quiet times) and their velocity to about 900 km s^{-1} (instead of 300). The region of modified wind travels out from the sun as a supersonic wave and when it reaches the magnetosphere it gives rise to several different phenomena which include

(*a*) a modification of the geomagnetic field, called a *magnetic storm*;

(*b*) increase in the intensity of the auroral luminosity at heights around 100 km;

(*c*) a change in the ionosphere, called an *ionospheric storm*.

When, as is frequent, storms of this kind occur in association with a solar flare, it is possible to estimate the time of travel of the particles by observing the time that elapses between the flare and the start of the terrestrial storm: it is usually about 36 hours.

It is convenient to divide the storm phenomena into those resulting from

(i) the direct effect of the enhanced solar wind, called DCF, 'disturbance corpuscular flux';

(ii) a modification of the ring current in the upper ionosphere; called DR, 'disturbance ring';

(iii) additional currents flowing in the polar ionosphere, called DP, 'disturbance polar current'.

(i) *The enhanced solar wind (DCF)*. On arrival at the boundary of the magnetosphere, the enhanced solar wind increases the compression of the geomagnetic field, already caused by the normal wind. On the simple hypothesis of a plane boundary between the solar wind and the geomagnetic field it was shown on p. 70 that for the normal wind with proton concentration $(n) = 5 \times 10^6\,\mathrm{m^{-3}}$ and velocity $(v) = 3 \times 10^5\,\mathrm{m\,s^{-1}}$ the boundary would be at a distance (d) of 10 earth radii and the field at the earth's surface would be increased by $\Delta B_R = 3 \times 10^{-9}\,\mathrm{T}$ (3 gamma). During a storm the wind increases suddenly so that $n = 10^7\,\mathrm{m^{-3}}$ and $v = 10^6\,\mathrm{m\,s^{-1}}$. A calculation like that on p. 70 shows that then $d = 6$ earth radii and $\Delta B_R = 10^{-8}\,\mathrm{T}$ (10 gamma). This increase of field occurs suddenly; it is called the *sudden commencement* of the storm. A more elaborate calculation, that makes a more reasonable assumption about the shape of the current sheet, leads to somewhat greater values of ΔB_R, more in accord with observation, but does not appreciably alter the magnitudes of the distances, d.

The enhanced wind, moving with a speed of about 1000 km s^{-1}, takes one or two minutes to envelop the sunward portion of the magnetosphere and the near portion of the magnetotail. The resulting field compressions are delayed as they travel towards the earth with the speed (about 1000 km s^{-1}) of hydromagnetic waves. The sudden commencement is thus observed, at different places on the earth, at times that may differ by as much as several minutes.

The sudden arrival of the enhanced solar wind has the form of a shock wave; it is followed by a more-or-less steady enhanced solar wind that results in a continuing compression of the geomagnetic field. The corresponding increase of field at the surface of the earth is called the *initial phase* of the storm (fig. 5.5).

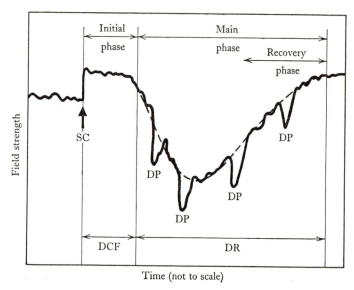

Fig. 5.5. Idealized storm variation of geomagnetic field. The part marked DCF is attributed to the impact of an enhanced solar wind (corpuscular flux) on the magnetosphere; that marked DR is attributed to an enhanced ring current; that marked DP is attributed to currents in the polar ionosphere.

(ii) *The ring current (DR)*. Shortly after the enhanced solar stream has reached the magnetopause there is an increase in the number of energetic particles in the magnetospheric trapping region. The particles, with energies of order 10 keV (electrons) or 10 MeV (protons), cannot simply be those of the enhanced wind which have energies of only a few eV (electrons) or a few keV (protons): their origin is not yet understood, it presents one of the outstanding problems of the magnetosphere.

As the more numerous energetic particles move in their trapped trajectories round the earth, electrons one way and protons the other, they correspond to an increased ring current and thus to a decrease of the magnetic field at the ground. This decrease is called the *main phase* of the magnetic storm; it is in the opposite sense to the increase that marks the sudden commencement and the initial phase. There is usually an interval of a few hours between the sudden commencement and the start of the main phase. The delay presumably represents the time taken for the ring current to build up. After new particles cease to be injected the ring current slowly decays for one or two days. During this time the excess particles are lost by some of the mechanisms discussed on p. 80, and the magnetic field gradually returns to its normal value: this is called the *recovery phase* of the magnetic storm.

The change ΔB of the geomagnetic field (B_0) at the surface of the earth is related to the energy W of the trapped particles by the expression

$$\Delta B/B_0 \fallingdotseq -W/W_{ext}$$

where W_{ext} represents the energy of the external magnetic field of the earth (p. 130). During magnetically quiet times the total energy of the trapped particles is thought to be about 10^{14}J, so that, with $W_{ext} = 10^{18}$J, $\Delta B/B_0 \fallingdotseq -10^{-4}$ representing a more or less constant depression of the field of order 3×10^{-9} T (3 gamma). During the main phase of a storm the energy in the ring current may increase by a factor of about 30, so that the field near the earth decreases by about 10^{-7} T (100 gamma).

(iii) *Additional currents in the polar ionosphere (DP)* [74, 114, 131, 133]. Sometimes comparatively short-lived changes occur in the polar magnetic field; they are called *polar substorms*. Although they are most frequent during the main and recovery phases of magnetic storms they are also observed during magnetically quiet times. They are of two kinds, called DP1 and DP2.

Those of type DP2 correspond to an increase in the twin current vortices that constitute the S_q^p current system of magnetically quiet times (fig. 5.3). The increase is presumably the result of increased convection in the magnetosphere resulting from the increased solar wind. Because the current vortices extend to middle latitudes this kind of substorm is noticeable outside the polar regions.

Substorms of type DP1 are associated with current systems that flow towards the west round ovals surrounding the two magnetic poles. These *auroral ovals* are closest to the poles on the sunward side and farthest removed in the anti-sun direction. The currents are most intense over a narrow region, the *auroral electrojet*, near the midnight part of the oval.

It is thought that the auroral oval corresponds to those magnetic field lines that leave the earth and stretch out into the magnetotail, and that the displacement of the oval towards the anti-sun direction corresponds to the distorted shape of the magnetosphere (see fig. 4.3). During disturbed times short-lived changes in the magnetotail, possibly resulting from field line reconnection, produce energetic particles that travel to the earth along lines of force, mostly into the midnight part of the oval. The particles penetrate to the low ionosphere, where they produce increased conductivity. The magnetospheric changes also produce electric fields that, transferred along the lines of force, drive the polar electrojet in the places where the conductivity is increased. The details of an explanation of this kind are still being worked out.

In recent times the interplanetary magnetic field has been measured during magnetic substorms observed on the earth's surface: the results have indicated that the storms often occur when the component of the interplanetary field that is parallel to the earth dipole axis changes its direction from north to south [131]. It is shown in §4.1 (p. 73) how a change of that kind might produce changes in the magnetosphere.

The energetic electrons deposited in the ionosphere during substorms produce aurorae which occur most frequently near the midnight part of the oval. As the earth rotates beneath the oval, observers at geomagnetic latitudes near 68 degrees come in succession below the midnight part: reports of aurorae thus come most frequently from places on a circle at this latitude: it is called the *auroral zone*.

As the electrons entering the ionosphere during a substorm penetrate to heights of 100 or 80 km they produce sufficient free electrons to reflect and absorb radio waves. The resulting layer of electrons in the E region is called *auroral sporadic E* or *auroral* E_s. Those electrons that penetrate to lower heights increase the absorption of radio waves, the phenomenon is then called *auroral absorption*; if it is so intense that radio waves used for communicating, or for ionospheric sounding, cannot be received, there is said to be an *auroral blackout*. The course of an auroral blackout can conveniently be followed by observing the radiation from a radio star, on a frequency so great that it is not completely absorbed in traversing the ionosphere. Observations of that kind are made with an instrument called a *riometer* (Relative ionosphere opacity METER).

5.2.3 Ionospheric storms (non-polar regions) [37, 58, 109, 113, 162]

In the polar regions the main effects of ionospheric storms are the increases in the electron concentrations in the E and D regions already mentioned: they can be understood in outline in terms of the energetic electrons that also produce the aurora. At lower latitudes, however, the changes in the D region take a different form and there are conspicuous changes in the F layer.

A storm usually results in significant changes in the F layer all over the world; at most latitudes the concentration of electrons is *decreased*, although, within a few degrees of the geomagnetic equator, it is often *increased*. These changes have not been properly explained. The decrease is the most surprising. One type of explanation ascribes it to an increase in the rate of loss of electrons. In the F region that loss involves reactions of the type $O^+ + N_2 \rightarrow N_2^+ + O$, with a rate $k[N_2][O^+]$: this rate could be increased during a storm either because k increases, or because $[N_2]$ increases. Laboratory measurements suggest that k increases with temperature and since the temperature is known to increase during a storm (fig. 1.13) the decrease of electron concentration would be explained.

Those who ascribe the storm phenomenon to an increase in the concentration of molecular nitrogen suggest that during a storm the entry of particles into the low polar ionosphere excites gravity or hydromagnetic waves that then travel into the F region at lower latitudes where they mix the atmospheric constituents.

Still another suggestion is that the modification of the ionospheric current system that shows itself as the polar electrojet is accompanied by an electric field that extends into the F region at lower latitudes and there causes the ionospheric plasma to move downwards to levels where the rate of loss is greater.

To explain an *increase* in the F region electron content, of the kind that is often observed near the equator during storms, it has been suggested that at those times the ionosphere is moved upwards to places where the rate of loss of electrons is smaller. The movement might be caused by the electric field of the dynamo below, so that the F region acts like an atmospheric motor. It might also be caused by a wind in

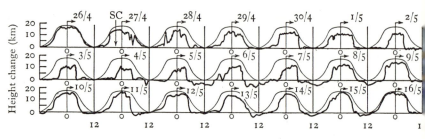

Fig. 5.6. Variations in the 'phase height' of waves of frequency 16 kHz observed at Cambridge on the dates shown in 1956. The thin line indicates the mean diurnal variation of phase for all the days of the month. A magnetic storm started with a 'sudden commencement' at the time marked SC on 26 April. From 0300 on 27 April to about 1200 on 28 April the phase was violently disturbed. During the nights from 30 April to 9 May the shape of the curve is unusual; after that it regains its normal shape slowly. This is the 'storm after-effect' (after reference 148).

the neutral air, blowing from the poles towards the equator, so as to move the ionization upwards along the sloping lines of force. Both these possibilities are being actively investigated at the time of writing.

Storm phenomena in the mid-latitude D region are most noticeable as changes in the phase of low-frequency radio waves reflected vertically from the region. In quiet times the phase varies smoothly and regularly throughout the day and night as shown by the thin line in fig. 5.6. During the initial and main phase of a storm the variations are irregular. During the recovery phase of the storm the diurnal change of radio phase is different from the usual; sudden changes near sunrise and sunset replace the normal smooth variation. The difference could be understood if, during this portion of the storm, electrons were continuously supplied in the lowest D region and were removed at night by attachment processes, but were liberated again by some detachment process during the day. On some occasions the abnormal type of phase variation has continued for 10 or 15 days after the storm, it has also been noticed at places far removed from the auroral region. These phenomena are not yet understood.

5.2.4 Proton storms. Polar cap events (PCEs) [43, 149]

The energetic streams of solar protons (accompanied by electrons) that move like independent particles behave in many ways like galactic cosmic rays: they can be called solar cosmic rays. Their energy is usually of order 10^6–19^9 eV: some of their important properties can be read from fig. 1.7. They travel from sun to earth in one or two hours. On arrival at the earth the protons are deviated by the geomagnetic field so that they impinge on the low ionosphere equally by day and by night but only within limited areas (the polar caps) around the geomagnetic poles. Unlike the particles that cause polar substorms, they impinge on the whole polar cap, not merely along the auroral oval. They produce ionization most copiously in the D region, at heights of 80–90 km: their presence is most clearly noticed through their effects on the strength of high-frequency radio waves and on the phase of low frequency waves.

Part 2

The principles governing some ionospheric processes and experimental methods

---

6 Collisions and diffusion

6.1 Collisions [44]

It is often necessary to consider how the movements of a set of charged particles denoted by (1) are altered when they make collisions with other charged, or neutral, particles, denoted by (2). It will be supposed that the particles are moving so slowly that no excitation or ionization occurs: kinetic energy is then conserved and the collisions are elastic. Suppose that the particles (1) move together with a bulk velocity V_1, superimposed on their much greater, and random, gas-kinetic velocities, so that to an observer moving with velocity V_1 their velocities appear to have a Maxwell distribution. Suppose also that the particles (2) have a Maxwell distribution of velocities with a bulk velocity V_2 superimposed, and for simplicity suppose V_1 and V_2 are both along the x-direction. When there are collisions between particles of the two groups we shall assume that the transfer of bulk momentum in the x-direction occurs on the average as though the random gas-kinetic velocities were absent. If a collision were head-on a particle (1) would thus gain momentum

$$\frac{2m_1 m_2}{m_1 + m_2}(V_2 - V_1)$$

and if it made, on the average, ν_{12} head-on collisions in unit time it would gain momentum at an average rate

$$\nu_{12}\left(\frac{2m_1 m_2}{m_1 + m_2}\right)(V_2 - V_1)$$

as though a force of the same magnitude acted upon it. Although the collisions are not, in reality, all head-on it is possible to describe the average rate of change of momentum in terms of an expression of this kind, and then ν_{12} is called the *collision frequency for momentum transfer*. Thus if there are n_1 particles in unit volume of (1) then the force F_{12} exerted on unit volume of (1) by the presence of (2) is given by

$$F_{12} = n_1 \nu_{12} \left(\frac{2m_1 m_2}{m_1 + m_2} \right) (V_2 - V_1) \tag{6.1}$$

and the force exerted on unit volume of (2) by the presence of (1) is given by the similar expression

$$F_{21} = n_2 \nu_{21} \left(\frac{2m_1 m_2}{m_1 + m_2} \right) (V_1 - V_2) \tag{6.2}$$

Since these forces must be equal and opposite

$$\nu_{12}/\nu_{21} = n_2/n_1 \tag{6.3}$$

The magnitude of ν under any particular circumstances is either calculated from the theory of the appropriate collisions, or is determined experimentally. The manner in which the factor $2m_1 m_2/(m_1 + m_2)$ is included in the result varies from worker to worker.

It will be supposed, following ordinary kinetic theory, that the probability of a collision that terminates a free path of duration between t and $t + dt$ is $\nu \exp(-\nu t)\, dt$; the average time between collisions is then $1/\nu$ as required by the definition of ν.

If the particles (1) are electrons (e) and (2) are electrically neutral atoms (n) with concentration n_n, an approximate estimate of ν_{en} can be derived on the supposition that the atom behaves like a sphere with cross-section σ and that the electron has a gas kinetic velocity $v = (3kT/m_e)^{\frac{1}{2}}$ then

$$\nu_{en} = n_n v \sigma = n_n (3kT/m_e)^{\frac{1}{2}} \sigma \tag{6.4}$$

To calculate an order of magnitude put $\sigma = \pi r^2$ where r is the atomic radius ($\fallingdotseq 10^{-10}$ m) to give

$$\nu_{en} \fallingdotseq 2 \times 10^{-16} \times n_n \times T^{\frac{1}{2}} \, \text{s}^{-1} \tag{6.5}$$

At a temperature of 900 °K this gives $\nu_{en} = 6 \times 10^{-15} \times n_n \, \text{s}^{-1}$.

A more thorough investigation shows that an electric dipole moment is induced in a neutral particle when a charged particle approaches it and that the effective cross-section for collision between them is determined to a large extent by the resulting electrostatic forces. At small enough temperatures the cross-section for an electron colliding with a neutral particle is proportional to $T^{\frac{1}{2}}$ so that the collision frequency is proportional to T or to the energy of the electron. This energy dependence of the collision frequency can be of some importance in calculations of the absorption of low-frequency radio waves.

When an ion collides with a neutral particle at small enough temperatures the force resulting from the induced moment leads to a cross-section that is proportional to $T^{-\frac{1}{2}}$, so that the collision frequency is independent of temperature: it is given by the expression

$$\nu_{\text{in}} = 2 \cdot 6 \times 10^{-15} \times n_{\text{n}} \times M^{-\frac{1}{2}} \, \text{s}^{-1} \tag{6.6}$$

where M is the mass of the ion (in atomic mass units).

When two charged particles collide the situation is quite different. If the collision is between an electron and a heavy ion, the ion can be considered to be almost stationary; the electric force then causes the electron to deviate from its path by an amount that depends on its speed and on the impact parameter. If the two charges were of the same sign the situation would be like that of an alpha particle approaching a much more massive positively charged nucleus: it was first investigated by Rutherford and is usually described as 'Rutherford scattering'. If an electron with velocity v makes a head-on approach to a singly-charged negative ion with charge e it is turned back on its tracks when it reaches a radial distance (r) where the electrostatic potential energy equals the original kinetic energy, so that

$$e^2/4\pi\epsilon_0 r = \tfrac{1}{2}mv^2$$

or

$$r = e^2/2\pi\epsilon_0 mv^2$$

If the approach is not head-on, but is such that the impact parameter is p, then the theory of Rutherford shows that when $p = 2r$ the particle is deviated through 90 degrees, so that all its forward momentum is lost. The radius for momentum loss is thus of order $e^2/\pi\epsilon_0 mv^2$, and the cross-section σ of order $(e^2/\epsilon_0 mv^2)^2$. This result is also applicable to the situation where the electron approaches a positive ion and the mutual force is one of attraction.

This collision radius is highly dependent on the velocity v, indeed if

v is small enough it can be larger than the mean distance between heavy particles. When that situation arises an electron's momentum is changed not only by a few large deviations as it approaches near to one or two ions, but also by several small deviations as it passes many ions at somewhat greater distances. When these small deviations are taken into account and when the previous calculation is carried out, with regard to the detailed statistics of the collision, the expression for the cross-section takes the form

$$\sigma = C(e^2/\epsilon_0 mv^2)^2 \tag{6.7}$$

where C is a factor that depends on the closeness of spacing of the particles and on their energies.

It is convenient to express (6.7) in terms of temperature by use of the relation $\frac{1}{2}mv^2 = \frac{3}{2}kT$ so that it becomes

$$\sigma = C(e^2/3\epsilon_0 kT)^2 \tag{6.8}$$

When detailed calculations are made [44] this leads to

$$\sigma = 6 \times 10^{-9} \times T^{-2} \, \text{m}^2 \tag{6.9}$$

It is interesting to compare the cross-section (σ_{ei}) for electron–ion collisions, with the cross-section (σ_{en}) for electron–neutral particle collisions, If we take $\sigma_{en} = 3 \times 10^{-20} \, (= \pi r^2)$ we have

$$\sigma_{ei}/\sigma_{en} = 2 \times 10^{11} \times T^{-2}$$

and if $250\,°\text{K} < T < 2500\,°\text{K}$ as in the ionosphere, this ratio varies from about 3×10^6 to 3×10^4, so that σ_{ei} is everywhere much greater than σ_{en}.

The collision frequency is proportional to $v\sigma$, where v is the relative velocity of the particles, proportional to $T^{\frac{1}{2}}$. Then, with $\sigma \propto T^{-2}$ from (6.8) the collision frequency is proportional to $T^{-\frac{3}{2}}$, and decreases as the temperature increases. In the ionosphere the light electrons have velocities much greater than the heavy ions so that the collision cross-section is determined by the temperature of the electrons: the fact that it decreases with increase of temperature can have important consequences discussed in § 3.3.

6.2 Diffusion

When several neutral gases rest in equilibrium under the action of gravity each is distributed as though it alone were present: thus if the temperature is uniform its concentration varies exponentially with a distribution height $H = kT/mg$. It is sometimes necessary to consider a situation in which the main gas (the major constituent) has this equilibrium height-distribution, while another gas (a minor constituent), is present with a much smaller concentration and with a distribution that is not its equilibrium one. The minor constituent then moves, through the major one, with a velocity that is determined by diffusion in the way described below.

Consider first a situation where there is no gravitational field and where the major gas (J) is at rest and is distributed uniformly. Let the minor constituent (N) be distributed so that the concentration (n) of its particles (mass m) has a gradient dn/dx along the x-direction, and let each of its particles make v collisions in unit time with particles of J. Let N drift with velocity W along the x-direction and let its partial pressure be p; since $p = nkT$ it is a function of x. Then each unit volume of N experiences forces

(*a*) $-nvmW$ resulting from the collisions between its particles and those of J.

(*b*) $-dp/dx$ resulting from its partial pressure gradient.

If the drift velocity W is constant, the total force is zero so that

$$-dp/dx = nmvW \tag{6.10}$$

or

$$-\frac{kT}{mv}\frac{dn}{dx} = nW$$

or

$$-D\frac{dn}{dx} = nW \tag{6.11}$$

where $D = kT/mv$ is called the diffusion coefficient.

Next suppose that x is measured vertically upwards and replace it by the symbol h, and suppose that the gravitational field of the earth is superimposed so that an additional force nmg acts downwards on each unit volume of the minor constituent. Equation (6.10) then becomes

$$-dp/dh - nmg = nmvW \tag{6.12}$$

and hence
$$-\frac{kT}{mv}\left(\frac{dn}{dh}+\frac{nmg}{kT}\right) = nW$$

or
$$-D\left(\frac{dn}{dh}+\frac{n}{H_N}\right) = nW \tag{6.13}$$

Here $H_N = kT/mg$, the scale height of the minor constituent. It is especially to be noticed that the scale height enters into (6.13) only as a constant, depending on the mass m and the temperature T, it is not necessarily the same as the distribution height $\left(-\frac{1}{n}\frac{dn}{dh}\right)$ of the minor constituent.

In discussing many phenomena in the ionosphere it is desirable to estimate the rate dn/dt at which the concentration of a minor constituent changes as a result of its diffusion through a background gas; it is equal to the difference between the rates at which particles enter a unit volume from below and leave it from above, so that

$$\frac{dn}{dt} = -\frac{d}{dh}(nW) = +\frac{d}{dh}\left\{D\left(\frac{dn}{dh}+\frac{n}{H_N}\right)\right\} \tag{6.14}$$

Before attempting to evaluate this expression we note that $D = kT/mv$ is inversely proportional to the frequency (v) of collisions between particles of the minor gas and of the major gas: it is thus inversely proportional to the concentration n_J of the major gas and can be written $D = D_0(n_{J0}/n_J)$ where D_0 and n_{J0} are values at some reference level. In most of the situations to be discussed here the major gas is in equilibrium in the gravitational field so that

$$n_J = n_{J0}\exp\left(-h/H_D\right)$$

where H_D is its scale height: then $D = D_0\exp\left(+h/H_D\right)$ and increases exponentially with height. With this height-dependence of D in mind, (6.14) can now be developed thus:

$$\frac{dn}{dt} = \frac{dD}{dh}\left(\frac{dn}{dh}+\frac{n}{H_N}\right)+D\left(\frac{d^2n}{dh^2}+\frac{1}{H_N}\frac{dn}{dh}\right)$$
$$= D\left\{\frac{d^2n}{dh^2}+\left(\frac{1}{H_D}+\frac{1}{H_N}\right)\frac{dn}{dh}+\frac{n}{H_N H_D}\right\} \tag{6.15}$$

The intermediate step in the algebra is included to emphasize the parts played by the concentration gradient $(\mathrm{d}n/\mathrm{d}h)$ of the minor constituent, the force of gravity on it (via H_N), and the height-distribution of the major constituent (via H_D).

It is next interesting to use (6.15) to determine how the magnitude of $\mathrm{d}n/\mathrm{d}t$ depends on the distribution of the minor constituent. For this purpose suppose that, at the particular height to be investigated, n varies like

$$n = n_0 \exp\left(-h/\delta\right) \tag{6.16}$$

so that, from (6.15),
$$\frac{\mathrm{d}n}{\mathrm{d}t} = \gamma n$$

with
$$\gamma = D\left(\frac{1}{\delta}-\frac{1}{H_D}\right)\left(\frac{1}{\delta}-\frac{1}{H_N}\right) \tag{6.17}$$

Thus, so long as the minor constituent remains distributed approximately exponentially with distribution height δ, its concentration n increases at the rate γn.

The magnitude of γ is discussed in appendix B where it is shown that, under most circumstances, it is of order $D/(H_N^2$ or H_D^2 or $\delta^2)$. This result can often be used to estimate the magnitude of $\mathrm{d}n/\mathrm{d}t$ resulting from diffusion and to compare it with the magnitude resulting from other processes, but in doing so it must be remembered that the distribution of the minor constituent does not usually retain the assumed exponential form for long.

6.2.1 Diffusion of plasma [118]

Next suppose that the minor constituent is a plasma of electrons and ions, with temperatures T_e and T_i respectively, and suppose that their concentrations (n) are equal. The electrons tend to diffuse more rapidly than the ions because of their smaller mass, but in doing so they set up a space charge and an electric field (E) that tends to slow them down. This field acts in the opposite direction on the ions and tends to speed them up, and we assume that a steady state is ultimately reached in which ions and electrons move with the same drift velocity and have the same concentrations, so that there is no volume space charge (the space charges responsible for the field E are supposed to reside on the boundaries of the volume considered and to be caused by the displacement of the electrons relative to the ions; they will be discussed shortly). The forces acting on the particles are their partial pressures, their weights, and the electric field E. The previous equation of motion (6.12) then becomes, for the ions and the electrons separately:

$$-\mathrm{d}p_i/\mathrm{d}h - nm_ig + Een = nWm_i\nu_i \qquad (6.18)$$

and

$$-\mathrm{d}p_e/\mathrm{d}h - nm_eg - Een = nWm_e\nu_e \qquad (6.19)$$

If we add these equations, suppose that $m_i \gg m_e$ and $m_i\nu_i \gg m_e\nu_e$ and write $p_i = nkT_i$ and $p_e = nkT_e$ we obtain

$$-nk(T_i + T_e)(\mathrm{d}n/\mathrm{d}h) - nm_ig = nWm_i\nu_i$$

or

$$-\frac{k(T_i + T_e)}{m_i\nu_i}\left\{\frac{\mathrm{d}n}{\mathrm{d}h} + \frac{nm_i\,g}{k(T_i + T_e)}\right\} = nW$$

or

$$-D_{\mathrm{amb}}\left\{\frac{\mathrm{d}n}{\mathrm{d}h} + \frac{n}{H_p}\right\} = nW \qquad (6.20)$$

where

$$D_{\mathrm{amb}} = \frac{k(T_i + T_e)}{m_i\nu_i} \qquad (6.21)$$

is called the *ambipolar diffusion coefficient* and

$$H_p = \frac{k(T_i + T_e)}{m_ig} \qquad (6.22)$$

is called the *plasma scale height*.

If electrons and ions have the same temperature T these expressions show that the ambipolar diffusion coefficient, and the plasma scale height, are each twice as great as the corresponding quantity for a gas of neutral particles having the same mass as the ions. The electrons have exerted their influence through the electric field so that the resulting velocities and the scale height are each doubled. It is interesting to examine the magnitude of this field a little more fully.

To understand the fundamentals of the matter, consider an equilibrium situation, where $W = 0$, suppose the two temperatures are equal, and neglect the electron mass compared with the ion mass: (6.18) then leads to

$$-kT\left(\frac{1}{n}\frac{\mathrm{d}n}{\mathrm{d}h}\right) - m_ig + Ee = 0 \qquad (6.23)$$

and (6.19) to
$$-kT\left(\frac{1}{n}\frac{dn}{dh}\right) - Ee = 0 \qquad (6.24)$$

Addition of (6.23) and (6.24) gives

$$\frac{1}{n}\left(\frac{dn}{dh}\right) = -\frac{m_1 g}{2kT} = -\frac{1}{H_p} \qquad (6.25)$$

and insertion of this into (6.24) gives

$$Ee = +\tfrac{1}{2}gm_1$$

The electric field thus acts downwards on the electrons with a force $\tfrac{1}{2}gm_1$ so that they behave as if they had mass $\tfrac{1}{2}m_1$: the same force acts upwards on the ions, and, when combined with the downwards force $m_1 g$ of gravity causes them, also, to behave as if their mass were $\tfrac{1}{2}m_1$.

The magnitude of the field $E = m_1 g/2e$ is quite small, only about $10^{-6}\,\mathrm{V\,m^{-1}}$ for oxygen atoms. This field might be produced by the displacement of a thin layer of electrons through a distance Δx beyond the top of a slab of ionosphere, leaving behind an equally thin layer of ions at the bottom. If the surface density of charge in the layers was $\pm\,\sigma$, then

$$E = (\sigma/\epsilon_0) = (ne/\epsilon_0)\,\Delta x \qquad (6.26)$$

Insertion of the value $E = \dfrac{kT}{ne}\dfrac{dn}{dh}$ from (6.24) yields

$$\Delta x = \left(\frac{\epsilon_0 kT}{ne^2}\right)\left(\frac{1}{n}\frac{dn}{dh}\right) = \lambda_D^2/\delta \qquad (6.27)$$

where $\lambda_D = (\epsilon_0 kT/ne^2)^{\frac{1}{2}}$ is the Debye length in the plasma (see (8.17)) and $\delta\left(=-\dfrac{1}{n}\dfrac{dn}{dh}\right)$ is the distribution height. In the ionosphere λ_D has values between about $10^{-2}\,\mathrm{m}$ and $10\,\mathrm{m}$, and δ is of order 10 to 50 km so that the required displacements would be very small, of order $10^{-3}\,\mathrm{m}$ or less.

6.2.2 Distribution of a plasma with several kinds of ions [49]

If more than one species of ion exists the upwards force of the electric field, being the same for all, combines with the different downwards forces of gravity to produce different effective masses for each species, for some the upwards force may even be greater than the downwards, so they may seem to have a negative mass and a negative distribution height. To examine this possibility let n_j and m_j be the concentration and the mass of the jth type of ion so that (6.23) and (6.24) become

for ions $\qquad -kT(dn_j/dh) - n_j m_j g + n_j Ee = 0 \qquad$ (6.28)

for electrons $\qquad -kT(dn_e/dh) \qquad\qquad - n_e Ee = 0 \qquad$ (6.29)

and, after addition:

$$-kT \frac{d}{dh}(\Sigma n_j + n_e) - \Sigma n_j m_j g + (\Sigma n_j - n_e) Ee = 0 \qquad (6.30)$$

Now note that electrical neutrality requires that $\Sigma n_j = n_e$ and write $\Sigma n_j m_j / n_e = \overline{m}_+$ to represent the mean ionic mass; (6.30) then becomes

$$\left. \begin{aligned} -2kT(dn_e/dh) - n_e \overline{m}_+ g &= 0 \\ \frac{1}{n_e} \frac{dn_e}{dh} &= -\frac{\overline{m}_+ g}{2kT} \end{aligned} \right\} \qquad (6.31)$$

or

The electrons thus have a distribution height equal to twice the scale height corresponding to the mean ion mass. From (6.29) and (6.31) $Ee = \frac{1}{2}\overline{m}_+ g$ and this upwards force acts on each ion in addition to the downwards force of gravity. The equation for the jth ion then becomes

$$-kT\left(\frac{1}{n_j}\frac{dn_j}{dh}\right) = (m_j - \tfrac{1}{2}\overline{m}_+)g \qquad (6.32)$$

so that it has a distribution height $kT/g(m_j - \frac{1}{2}\overline{m}_+)$ and if $\frac{1}{2}\overline{m}_+ > m_j$ this is negative. It should be noticed that $\overline{m}_+$ depends on the relative numbers of the different ions so that it is height-dependent: thus (6.32) does *not* indicate that any one type of ion is distributed exponentially. A solution of the simultaneous differential equations for the situation where O^+ H^+ and He^+ ions are present with the same uniform temperature is represented in fig. 6.1.

6.2.3   Diffusion affected by the geomagnetic field

The presence of a magnetic field can have a profound influence on the diffusion of a plasma: it is sufficient to consider only the motion of the ions since the electrons are forced to follow them. Suppose the magnetic field B is along the z-direction and that unit forces act on unit volume in the x- and the z-directions. The resulting motion is discussed in §7.2.1 where (7.17) (7.18) (7.19) and (7.20) show that the resulting velocities V_x, V_y, and V_z along the x- y- and z-directions are proportional to $r(1+r^2)^{-1}$, $-(1+r^2)^{-1}$ and r^{-1}, where $r = v/\Omega$. The important

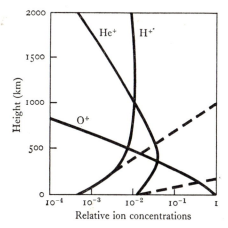

Fig. 6.1. To illustrate a possible height-distribution of ions in a mixture of H^+, He^+, O^+ and electrons. The temperature is taken to be 1200 °K and the relative concentrations at zero height are assumed to be as shown. The dashed lines correspond to the conditions of photochemical equilibrium described in §3.2 (after reference 49).

situations arise at the greater heights where $r \ll 1$ and there the ratios of the velocities are:

$$|V_y/V_z| = r/(1+r^2) \ll 1$$

$$|V_x/V_z| = r^2/(1+r^2) \ll 1$$

The velocity along the field is thus much more important than that across the field.

Now consider the situation previously discussed, where gravity and a partial pressure gradient together exert a vertical force F on each unit volume, in the presence of the earth's magnetic field inclined at an angle I to the horizontal, and at a height where $v_i \ll \Omega_i$. Only the component $F \sin I$ of the force along the field produces any important velocity, and diffusion occurs almost entirely in the field direction, with a velocity proportional to $\sin I$. The component of this velocity in the vertical direction is again smaller by a factor $\sin I$ so that the velocity W in the vertical direction is given, not by (6.13), but by

$$nW = -D\sin^2 I \left(\frac{dn}{dh} + \frac{n}{H_N} \right) \tag{6.33}$$

and the rate of change of concentration is given by

$$\frac{dn}{dt} = D\sin^2 I \left\{ \frac{d^2 n}{dh^2} + \left(\frac{1}{H_D} + \frac{1}{H_N} \right) \frac{dn}{dh} + \frac{n}{H_D H_N} \right\} \tag{6.34}$$

instead of by (6.15).

Direct application of these equations would lead to the conclusion that diffusion in the vertical direction is much slowed down near the equator where the magnetic field is nearly horizontal. There are, however, complicating factors when the motion of the plasma is communicated to the neutral air: they are discussed below.

6.3   Air drag and ion drag [72, 174]

If the ionospheric plasma and the neutral air are moving relative to each other collisions between their particles can alter their bulk motions. It is said that the air exerts an *air drag* on the ionosphere and the ionosphere exerts an *ion drag* on the air. Consider a situation in which the plasma moves with bulk velocity V_p and the neutral air with velocity V_a in the same direction so that collisions between the ions of the plasma and the molecules of the air tend to equalize their velocities. A unit volume of plasma and a unit volume of air experience equal and opposite forces of magnitude, (p. 102),

$$F_p = \nu_{pa} n_p \left(\frac{2m_p m_a}{m_p + m_a} \right) (V_a - V_p) \tag{6.35}$$

$$F_a = \nu_{ap} n_a \left(\frac{2m_p m_a}{m_p + m_a} \right) (V_p - V_a) \tag{6.36}$$

which produce accelerations

$$dV_p/dt = F_p/n_p m_p \fallingdotseq \nu_{pa}(V_a - V_p) \qquad (6.37)$$

$$dV_a/dt = F_a/n_a m_a \fallingdotseq \nu_{ap}(V_p - V_a) \qquad (6.38)$$

Now suppose that the plasma is kept moving at a constant velocity V_p by an imposed electric field: and that initially the air is stationary. Integration of (6.38) shows that, if the air is acted on by no other forces its velocity will increase like

$$V_a = V_p\{1 - \exp(-\nu_{ap} t)\} \qquad (6.39)$$

with a time constant ν_{ap}^{-1} until its velocity is the same as the plasma's. The same kind of argument shows that if outside forces cause the air to move with velocity V_a the velocity of the plasma will approach that velocity with a time constant ν_{pa}^{-1}. Rough values for these time constants are shown in table 5. Air drag thus causes the ionosphere to follow the air movements with a time lag of one second or less, and ion drag causes the air to follow the movements of the plasma with a time lag of order one hour.

TABLE 5. *Time constants for 'air drag'* (ν_{pa}^{-1}) *and 'ion drag'* (ν_{ap}^{-1})

height (km)	$\nu_{ap}^{-1}(h)$	$\nu_{pa}^{-1}(s)$
100	3	10^{-4}
200	1.5	0.25
300	0.5	2

In table 5 it is noticeable that, as the height increases, ν_{ap}^{-1} decreases while ν_{pa}^{-1} increases. The reason for this difference is that ν_{ap} is proportional to the concentration of the ions, which is greater at the greater heights, but ν_{pa} is proportional to the concentration of the neutral air particles, which is smaller at the greater heights.

In the foregoing discussion of ion drag it was supposed that the velocity of the plasma was maintained constant and that no outside force acted on the air. In some important situations, however, the movement of the plasma is the result of diffusion, and the force of gravity acts on the air. Consider for example a situation where the diffusion of the ionosphere by itself would be along sloping lines of force: it would try to set the air in motion in the same direction with time constant ν_{ap}^{-1}, but since the air is prevented, by gravitational forces, from moving in the vertical direction, it finally assumes a horizontal velocity equal to the horizontal projection of the plasma's velocity. When the interplay of forces is considered in detail it is found that the diffusion then occurs purely in the vertical direction just as it would if there were no magnetic field. The time constant for the production of this situation is ν_{ap}^{-1}: reference to table 5 shows that it is of order one hour. (See p. 141 in ref. 28.)

7 Movements of charged particles in magnetic fields†

7.1 Movements without collisions

7.1.1 Motion of a free charged particle

Consider the motion of a particle of mass m and charge e that makes no collisions and that moves freely in the presence of the earth's magnetic field B. Axes are chosen so that B is along Oz and the equations of motion are

$$m\dot{V}_x = eBV_y \tag{7.1}$$

$$m\dot{V}_y = -eBV_x \tag{7.2}$$

$$m\dot{V}_z = 0 \tag{7.3}$$

If the initial conditions are $V_x = V_{x0}$, $V_z = V_{z0}$, $V_y = 0$ these equations lead to the solution

$$V_x = V_{x0} \cos \Omega' t \tag{7.4}$$

$$V_y = -V_{x0} \sin \Omega' t \tag{7.5}$$

$$V_z = V_{z0} \tag{7.6}$$

where

$$\Omega' = eB/m \tag{7.7a}$$

The particle thus travels along the z-direction with velocity unaltered and moves in a circle around that direction with the angular gyro-frequency Ω' so that the resultant motion has a helical form. Since the sign of Ω' depends on the sign of e, positive and negative particles rotate in opposite directions.

Throughout the rest of this book the gyro-frequency (Ω) is defined so that it is independnet of the sign of the charge, by writing

$$\Omega = |e| B/m \tag{7.7b}$$

In the ionosphere the angular gyro-frequencies (which will be called the gyro-frequencies for short) of electrons and ions decrease with increasing height as the earth's magnetic field decreases: for ions there is an additional change as the mass of the predominant ion changes

† [8, 25, 84].

with height. Approximate magnitudes for electrons and ions are shown in table 6.

TABLE 6

Height	250 km	2000 km	2 earth radii	5 earth radii	Solar wind
Ω_e (s^{-1})	8.5×10^6	4×10^6	10^6	5×10^4	500
Ω_i (s^{-1})	250	50	500	30	0.3

7.1.2 Motion under a constant applied force

Next consider the motion of a free positively charged particle under the action of a constant force with components F_x and F_z. The equations of motion are then

$$m\dot{V_x} = F_x + eBV_y \tag{7.8}$$

$$m\dot{V_y} = \quad -eBV_x \tag{7.9}$$

$$m\dot{V_z} = F_z \tag{7.10}$$

The motions in the *x*- and *y*-directions can be discussed as usual in terms of a particular integral (which may be any solution that satisfies (7.8) and (7.9)) together with a complementary function (which is the solution of (7.8) and (7.9) when $F_x = 0$). The complementary function must be chosen so as to satisfy the initial conditions of the problem. A particular integral is $V_x = 0$, $V_y = -F_x/eB$: if this happens to be the initial velocity then no complementary function is needed, the motion continues as a constant velocity along the negative *y*-direction and all conditions are fulfilled.

If, however, at time $t = 0$, the particle is moving with velocity (V_{x0}, V_{y0}, V_{z0}) then a complementary function must be added; in the *xy* plane it has the form of the circular motion described by (7.4) and (7.5) and must be such that, when combined with the particular integral, it satisfies the initial conditions. The final solution is then

$$V_x = V_{x0} \cos \Omega t + (V_{y0} + F_x/eB) \sin \Omega t \tag{7.11}$$

$$V_y = (V_{y0} + F_x/eB) \cos \Omega t - V_{x0} \sin \Omega t - F_x/eB \tag{7.12}$$

$$V_z = V_{z0} + (F_z/m) t \tag{7.13}$$

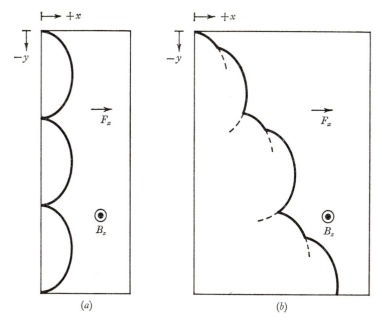

Fig. 7.1. If a magnetic field B_z is directed out of the plane of the paper a positively charged particle initially at rest, and acted on by a mechanical force F_x, follows the cycloidal path shown at (a). The resulting drift motion is in a direction perpendicular to the force. If the particle described in (a) makes collisions at random times, and starts with zero velocity after each collision, it follows the path shown at (b). There is a steady drift in the direction of the force, and the drift velocity perpendicular to the force is altered.

For a positive particle it represents a motion in a circle that drifts steadily along the negative y-direction; for a negative particle the drift is along the postive direction. The speed of the drift is $F_x/|e|B$, the same for electrons and for positive ions. Fig. 7.1 (a) shows the path of a positive particle that starts with $V_{x0} = V_{y0} = 0$: it is a cycloid. The motion in the z-direction proceeds independently as given by (7.13).

7.2 Movements in the presence of collisions

7.2.1 Motion under a constant applied force

If a charged particle makes collisions at random times, on the average $1/\nu$ apart, and if a constant force F_x acts on it in the presence of a magnetic field B_z, the drifting circular motion described in §7.1.2 is interrupted. After each interruption the ordered momentum is lost and the motion starts again in a random direction. The situation can be represented in a simple diagram if it is supposed that the particle starts from rest after each collision; fig. 7.1 (*b*) then indicates the type of path followed, it should be compared with (*a*) that shows the path followed by a particle that starts from rest in the absence of collisions. The effect of collisions is to introduce a component of steady drift in the direction (*Ox*) of the applied force and to modify the drift in the perpendicular (*Oy*) direction.

When many particles are present it is possible to calculate their average velocity in the *x*- and *y*- directions as follows. Assume that the probability of a collision's terminating a free path of duration between t and $t + \mathrm{d}t$ is $\nu e^{-\nu t} \mathrm{d}t$. The average velocity is obtained by adding the contributions to V_x and V_y during the intervals of the particles' free travels and dividing by the total number of particles, on the assumption that the initial velocities V_{x0} and V_{y0} after each collision are randomly distributed. From (7.11) and (7.12) we then have, for a particle with positive charge,

$$\overline{V}_x = \frac{\int_0^\infty \nu e^{-\nu t}(F_x/eB)\sin\Omega t\,.\,\mathrm{d}t}{\int_0^\infty \nu e^{-\nu t}\,.\,\mathrm{d}t} = \frac{F_x}{eB}\frac{\Omega\nu}{\Omega^2+\nu^2} \qquad (7.14)$$

$$\overline{V}_y = \frac{\int_0^\infty \nu e^{-\nu t}(F_x/eB)(\cos\Omega t - 1)\,\mathrm{d}t}{\int_0^\infty \nu e^{-\nu t}\,.\,\mathrm{d}t} = -\frac{F_x}{eB}\frac{\Omega^2}{\Omega^2+\nu^2} \qquad (7.15)$$

$$\overline{V}_z = \frac{\int_0^\infty \nu e^{-\nu t}(F_z/m)\,t\,.\,\mathrm{d}t}{\int_0^\infty \nu e^{-\nu t}\,.\,\mathrm{d}t} = \frac{F_z}{m\nu} \qquad (7.16)$$

Since (7.14), (7.15) and (7.16) are required for application to ions with positive charge and to electrons with negative charge, it is convenient to use subscripts i and e to denote ions and electrons respectively and to write $r = \nu/\Omega$ and to define functions

$$f_0(r) = \frac{1}{r} \qquad f_1(r) = \frac{r}{1+r^2} \qquad f_2(r) = \frac{1}{1+r^2} \qquad (7.17)$$

The functions are numbered so that they correspond to quantities σ_0, σ_1, σ_2 derived below. By using (7.17) the expressions (7.14, 7.15, 7.16) can then be usefully written

	positive ions	electrons	
$\overline{V}_x = +f_1(r_i)$	$\times \dfrac{F_x}{|e|B}$	$+f_1(r_e)$ $\times \dfrac{F_x}{|e|B}$	(7.18)

$$\overline{V}_x = +f_1(r_i) \Big| \times \frac{F_x}{|e|B} \quad\quad +f_1(r_e) \Big| \times \frac{F_x}{|e|B} \quad\quad (7.18)$$
$$\overline{V}_y = -f_2(r_i) \Big| \quad\quad\quad\quad +f_2(r_e) \Big| \quad\quad\quad\quad\quad (7.19)$$

$$\overline{V}_z = +f_0(r_i) \times \frac{F_z}{|e|B} \quad\quad +f_0(r_e) \times \frac{F_z}{|e|B} \quad\quad (7.20)$$

Frequently an electric field (E_x) is responsible for producing the force $F_x = eE_x$ and then the expressions become

positive ions | electrons

$$\overline{V}_x = +f_1(r_i) \Big| \times \frac{E_x}{B} \quad\quad -f_1(r_e) \Big| \times \frac{E_x}{B} \quad\quad (7.21)$$
$$\overline{V}_y = -f_2(r_i) \Big| \quad\quad\quad\quad -f_2(r_e) \Big| \quad\quad\quad\quad (7.22)$$

$$\overline{V}_z = +f_0(r_1) \times \frac{E_z}{B} \quad\quad -f_0(r_e) \times \frac{E_z}{B} \quad\quad (7.23)$$

Sometimes attention is directed to the current densities

$$j_x, j_y, j_z \,[= ne(V_x, V_y, V_z)]$$

and the associated conductivities $\sigma_1 = j_x/E_x$, $\sigma_2 = j_y/E_x$, and $\sigma_0 = j_z/E_z$ and then we can write

positive ions | electrons

$$\sigma_1 = +f_1(r_i) \Big) \quad\quad +f_1(r_e) \Big) \quad\quad\quad (7.24)$$
$$\sigma_2 = -f_2(r_i) \Big\} \times \frac{n\,|e|}{B} \quad +f_2(r_e) \Big\} \times \frac{n\,|e|}{B} \quad (7.25)$$
$$\sigma_0 = +f_0(r_i) \Big/ \quad\quad +f_0(r_e) \Big/ \quad\quad\quad (7.26)$$

It is useful to discuss the magnitudes of the velocities and currents at different heights in an atmosphere at uniform temperature in which $\nu = \nu_0 \exp(-h/H)$ so that $r(= \nu/\Omega) = r_0 \exp(-h/H)$. The functions $f_0(r)$, $f_1(r)$ and $f_2(r)$ are plotted on a logarithmic scale in fig. 7.2: they can be used to discuss the situation that arises in the ionosphere.

Two sets of curves are plotted, one above to represent $f(r_i)$ for ions,

5

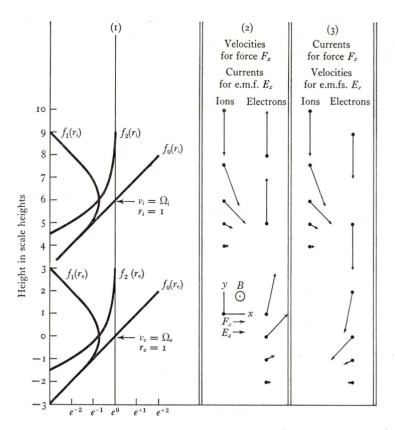

Fig. 7.2. *Column* (*1*) shows the functions

$$f_0(r) = \frac{1}{r}, \quad f_1(r) = \frac{r}{1+r^2}, \quad f_2(r) = \frac{1}{1+r_2},$$

with $r = r_0 \exp(-/hH)$ plotted against h/H for electrons below and for ions above. The curves for electrons are centred on the height where

$$\nu_e = \Omega_e \quad (r_e = 1)$$

and for ions where $\nu_i = \Omega_i$ ($r_i = 1$): they are shown separated by $6H$ to correspond roughly to the situation in the ionosphere. If $r_e = \nu_e/\Omega_e$ and $r_1 = \nu_i/\Omega_i$ they can be used, with the signs shown in 7.18–7.26 to represent the velocities, currents, and conductivities for electrons and ions respectively.

Columns (*2*) *and* (*3*) represent the velocities, of ions and electrons, and the currents that they produce, at different heights when a mechanical force F_x, or an electric field E_x acts. The plane xy is in the paper and a magnetic field B_z is supposed to be applied perpendicular to it.

and one below to represent $f(r_e)$ for electrons. The origins on the two height scales are at levels where $\nu = \Omega$, $(r = 1)$: for electrons this is near 80 km and for ions near 140 km, about $6H$ higher up. For electrons the curves $f_1(r)$, $f_2(r)$ and $f_0(r)$ describe the velocities in the x-, y-, z-directions, in terms of $(B\,|e|)^{-1}$ as a unit, when unit forces act in the x- and the z-directions. For positive ions the corresponding velocities are $f_1(r)$, $-f_2(r)$, and $f_0(r)$. The sketches in column 2 represent the resulting velocities in the xy plane when a mechanical force F_x acts in the x-direction. The following results are of interest.

1. At all heights the velocity (V_z) along the field is greater than the velocity in other directions, and above the level where $\nu = \Omega$ it rapidly becomes much greater. At all heights the velocity (V_z) of electrons is much greater than that of ions.

2. Low down, where ν/Ω is large, ions and electrons both move in the direction of the applied force: the velocities are smaller at the smaller heights where the collisions are more numerous.

3. High up, where ν/Ω is small, a force perpendicular to the magnetic field produces movements of ions and electrons perpendicular to itself but in opposite senses: the velocities reach a limiting value at heights greater than about $2H$ above the zero levels. The movements in opposite senses constitute a current, thus a mechanical force perpendicular to the magnetic field produces a current perpendicular to itself and to the field.

4. At the levels where $\nu = \Omega$ the velocities perpendicular to the magnetic field make angles of ± 45 degrees with the direction of the force.

Two important deductions follow from these results when the force that produces the movement arises from a wind in the neutral air. First, at levels where $\nu \ll \Omega$ a wind produces much greater velocities if it is along the direction of the magnetic field than if it is perpendicular to it, and second, a force perpendicular to the field does not necessarily produce movements along its own direction.

If the movements of electrons and positive ions are caused by an electric field E_x, equations (7.21), (7.22), show how the velocities in the xy plane are related to the functions f_1, f_2; they are sketched in column 3 of fig. 7.2. The most important point is that at heights exceeding the level where $\nu_i = \Omega_i$ by more than one scale height (i.e.

about 150 km) ions and electrons move together, in a direction perpendicular to the electric and the magnetic fields, with a velocity $V_y = -E_x/B$: the plasma thus moves bodily. If E and B are the vectors representing the total fields, the velocity can be represented vectorially by $V = (E \times B)/|B|^2$.

It is particularly to be noticed that at heights in the F region where

$$\nu_i/\Omega_i \ll 1 \quad \text{and} \quad \nu_e/\Omega_e \ll 1$$

a *mechanical force* normal to B produces a *current* normal to itself,

whereas

an *electric field* normal to B produces a *bodily movement* normal to itself.

7.2.2 Movement of an ionospheric irregularity [87, 96]

The previous discussion has shown that if the uniform ionosphere is acted on by an electric field at heights greater than about 130 km, where $\nu_i < \Omega_i$, it moves bodily, but that no current flows in it. If a limited 'cloud' of increased plasma concentration is situated in it space charges appear on the surface of the cloud and produce electric fields inside and outside the cloud. The field inside drives a current across the magnetic field: this current flows out at one side of the cloud into the surrounding ionosphere and back into it at the other side. In general the cloud changes its shape and moves: the change of shape is additional to any change that is produced by diffusion. The bodily movement of the cloud can be calculated if the currents that flow through it and through the rest of the ionosphere are known. The current flows through the ionosphere much more easily along the magnetic field lines at heights greater than 130 km ($\nu_i < \Omega_i$) whereas at smaller heights (where $\nu_e > \Omega_e$) it can equally easily flow across the field (see §7.2.3). The current that flows from the cloud into the ionosphere thus follows field lines from the cloud down into the E region, there it crosses the field and flows up along other field lines to complete the circuit.

To calculate the movement of the cloud it is necessary to calculate the polarization charges on its surface and the currents that they drive through the cloud and through the loop in the surrounding plasma that extends down into the E region. The problem is complicated and it is not easy to calculate the cloud's movements in terms of the originally imposed electric field.

In spite of the difficulties encountered in the calculations useful attempts have been made to determine the electric field in the ionosphere by observing the motion of ionized clouds produced artificially (§ 10.6). It might be expected that the current that flows across the magnetic field in the E region below the cloud would produce observable modifications of that region: experiments seem to confirm that expectation [159].

7.2.3 Ionospheric conductivities [67]

When discussing the currents that flow in the ionosphere it is convenient to make use of the conductivities given by (7.24), (7.25) and (7.26). Since these are proportional to the concentration of the electrons and ions it is best in a general discussion to consider the conductivities per ion pair, in terms of $|e|/B$ as a unit: the conductivities in the ionosphere are then derived by multiplying by the appropriate concentrations. The ionospheric current is obtained by adding the separate currents for ions and electrons: column (2) of fig. 7.2 shows that the currents in the direction of E_x (corresponding to σ_1) add, whereas those in the direction perpendicular to E_x (corresponding to σ_2) subtract. The resulting height variations of σ_0, σ_1, and σ_2 (per ion pair) in an idealized situation where $\nu = \nu_0 \exp(-h/H)$ are shown in fig. 7.3: in the actual ionosphere the levels marked 0 and 6 are near 80 and 140 km respectively. The specific conductivities are obtained by multiplying the conductivities per ion pair by the electron concentration, which increases with height, and by making proper allowance for height-variations of temperature and the corresponding variations of ν.

When an electric field is applied normal to a magnetic field, σ_1 corresponds to the current that flows in the direction of the electric field; it is called the *Pedersen conductivity*. σ_2 corresponds to the current that flows in a direction normal to both the electric and the magnetic fields; it is called the *Hall conductivity*. When an electric field is applied along the direction of a magnetic field, σ_0 corresponds to the

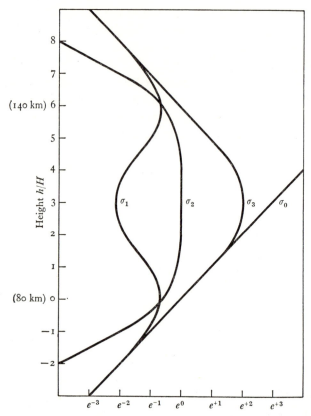

Fig. 7.3. Conductivities per ion pair in units of $|e|/B$ in a model ionosphere where $\nu = \nu_0 \exp(-h/H)$: $\nu_e = \Omega_e$ at $h = 0$, $\nu_i = \Omega_i$ at $h = 6H$: and B is the earth's magnetic field. In the real atmosphere the places where $h/H = 0$ and 6 are at heights of about 80 and 140 km.

σ_0 refers to the current that flows in the direction of an electric field applied parallel to the magnetic field (direct conductivity).

For the other conductivities the electric field is applied perpendicular to the magnetic field and

σ_1 refers to the current that flows in the direction of the electric field (Pedersen conductivity);

σ_2 refers to the current that flows perpendicular to the electric field and to the magnetic field (Hall conductivity);

σ_3 refers to the current that flows in the direction of the electric field if flow of current at right-angles is prevented (Cowling conductivity).

current that flows along the direction of the field; it is the same as the conductivity in the absence of a magnetic field.

If an electric field with components E_x, E_y, E_z is applied and if B is along the z-direction as before, the three conductivities can be used to express the result in terms of a tensor conductivity as follows

$$\begin{vmatrix} j_x \\ j_y \\ j_z \end{vmatrix} = \begin{vmatrix} E_x \\ E_y \\ E_z \end{vmatrix} \times \begin{vmatrix} \sigma_1 & -\sigma_2 & 0 \\ \sigma_2 & \sigma_1 & 0 \\ 0 & 0 & \sigma_0 \end{vmatrix} \qquad (7.27)$$

The phenomena so far discussed occur in an infinite unbounded assembly of electrons and ions: the ionosphere is, however, bounded above and below and the situation is different. To examine it suppose that the plasma previously discussed is bounded by two parallel planes perpendicular to Oy, and that a field E_x is applied. A current (j_y) then flows and builds up charges on the boundary planes until they produce an additional field E_y just sufficient to prevent any further current j_y; this field also affects the current in the Ox direction. The situation is then represented by the equations

$$j_x = \sigma_1 E_x + \sigma_2 E_y \qquad (7.28)$$

$$j_y = -\sigma_2 E_x + \sigma_1 E_y = 0 \qquad (7.29)$$

therefore
$$E_y = (\sigma_2/\sigma_1) E_x \qquad (7.30)$$

and
$$j_x = (\sigma_1 + \sigma_2^2/\sigma_1) E_x \equiv \sigma_3 E_x \qquad (7.31)$$

The conductivity σ_3 describes the current that flows along the direction of the electric field when current in the perpendicular direction is inhibited: it is sometimes called the *Cowling conductivity*.

If only one kind of charged particle were free to move the magnitudes of σ_1 and σ_2 would be given by (7.24) and (7.25) and insertion of these into (7.31) shows that $\sigma_3 = \sigma_0$. It follows that in a bounded medium containing only one kind of moving charge, currents flow just as they would if there were no magnetic field. This is what happens when current flows in a bounded piece of metal: although the superposition of a magnetic field does not alter the conductivity, it nevertheless produces an electrostatic field (E_y) perpendicular to the boundaries; this is the Hall effect.

The situation is, however, different in the ionosphere, where both ions and electrons can move; σ_1 and σ_2 are then each the sum of ion and electron conductivities, and σ_3 is no longer equal to σ_0. The magnitude of σ_3, per ion pair, in the model ionosphere is plotted, as a function of height, in fig. 7.3.

Now take Oy in the vertical direction and consider the ionosphere, containing electrons and positive ions both capable of moving, bounded top and bottom by horizontal planes perpendicular to Oy. Consider also the situation at the equator, where the earth's magnetic field (B_z) is horizontal, and suppose that a horizontal electric field E_x is applied perpendicular to B_z. A current then flows in the x-direction and the curve for σ_3 (fig. 7.3) shows that it is greatest at heights near 110 km, and is considerably greater than the currents that would flow in an unbounded slab. In a simplified form of the atmospheric dynamo theory this is supposed to be the situation at the geomagnetic equator.

Over other parts of the earth the direction of the magnetic field is inclined to the top and bottom boundaries of the ionospheric slab and the situation is a little more complicated. The current flowing in the horizontal, xz, plane is then determined by a tensor conductivity

$$\boldsymbol{\sigma} = \begin{vmatrix} \sigma_{xx} & \sigma_{xz} \\ \sigma_{zx} & \sigma_{zz} \end{vmatrix} \tag{7.32}$$

If the magnetic field is in the yz plane, and makes an angle I with the horizontal (Oz) it can be shown that

$$\sigma_{xx} = \frac{\sigma_0 \sigma_1}{\sigma_0 \sin^2 I + \sigma_1 \cos^2 I} \tag{7.33}$$

$$\sigma_{xz} = -\sigma_{zx} = \frac{\sigma_0 \sigma_2 \sin I}{\sigma_0 \sin^2 I + \sigma_1 \cos^2 I} \tag{7.34}$$

$$\sigma_{zz} = \frac{\sigma_2 \cos^2 I}{\sigma_0 \sin^2 I + \sigma_1 \cos^2 I} + \sigma_1 \tag{7.35}$$

Since in general $\sigma_0 \gg \sigma_1$ and $\gg \sigma_2$ these expressions can be written approximately

$$\sigma_{xx} \fallingdotseq \sigma_1/\sin^2 I \tag{7.36}$$

$$\sigma_{xz} \fallingdotseq \sigma_2/\sin I \tag{7.37}$$

$$\sigma_{zz} \fallingdotseq \sigma_1 \tag{7.38}$$

except when $\sigma_0 \sin^2 I \ll \sigma_1 \cos^2 I$ (or $\tan^2 I \ll \sigma_1/\sigma_0$)

and then

$$\sigma_{xx} \fallingdotseq \sigma_0 \qquad (7.39)$$

$$\sigma_{xz} \fallingdotseq 0 \qquad (7.40)$$

$$\sigma_{zz} \fallingdotseq \sigma_2^2/\sigma_1 + \sigma_1 \qquad (7.41)$$

The conductivity expressed by (7.39), (7.40) and (7.41) has the magnitude previously deduced for the special situation where the magnetic field is horizontal. Comparison with (7.36), (7.37) and (7.38) shows that it is much greater than the conductivity at places where it cannot be assumed that $\tan^2 I \ll \sigma_1/\sigma_0$. An enhanced conductivity of this kind near the equator in a slab-like ionosphere is supposed to be responsible for a relatively large current in the atmospheric dynamo, known as the *equatorial electrojet*.

7.3 Converging field lines. Trapped particles

The electrons and ions in the highest part of the atmosphere make a negligible number of collisions and move in helices along the geomagnetic field lines. The field is, however, not uniform; it decreases radially outwards and is represented by field lines that are curved and that become closer together as they approach the poles. Each of these variations produces its own modification of the simple helical motion that occurs in a uniform field. The modifications can be described by saying that the particle moves in a circle round a *guiding centre* which itself moves in a way determined by the shape of the field.

Suppose that at $z = 0$ the field B is along the z-direction and has magnitude B_0: a charged particle moving with velocity $v_\perp$ perpendicular to the z-direction then moves in a circle with radius $r = mv_\perp/Be$. It is equivalent to a current $(ev/2\pi r)$ enclosing an area πr^2, and hence has a magnetic moment

$$\mu = \tfrac{1}{2}mv_\perp^2/B_0 = W_\perp/B_0 \qquad (7.42)$$

where $W_\perp$ is the kinetic energy of the particle's motion in its circular orbit.

If the particle also has a velocity $v_\parallel$ along the z-direction it moves in a helix along the field and reaches a place where the field has a magnitude B different from B_0. It can be shown as follows that during this motion the velocity $v_\perp$ changes so that the magnetic moment (μ) remains constant with the magnitude given by (7.42).

If the field changes at a rate $\partial B/\partial t$ there is an e.m.f. $\pi r^2(\partial B/\partial t)$ acting round the orbit of the particle, so that in one revolution the kinetic energy of the particle is changed by

$$\Delta W_\perp = e\pi r^2(\partial B/\partial t) \tag{7.43}$$

If ΔB is the change in B during the time $(2\pi r/v_\perp)$ of one revolution then

$$\frac{\partial B}{\partial t} = \frac{v_\perp \Delta B}{2\pi r} \tag{7.44}$$

and, after substitution for r and $\partial B/\partial t$ into (7.43)

$$\Delta W_\perp = \tfrac{1}{2}mv_\perp^2 \Delta B/B = W_\perp(\Delta B/B) \tag{7.45}$$

Now the magnetic moment changes by $\Delta\mu$ during one revolution, where

$$\Delta\mu = \Delta\left(\frac{W_\perp}{B}\right) = \frac{1}{B}\Delta W_\perp - \frac{W_\perp}{B^2}\Delta B \tag{7.46}$$

and (7.45) shows that this is zero. It follows that the magnetic moment continues to have the magnitude given by (7.42).

Now consider how the convergence of the field lines affects the particle's motion. The convergence is a pictorial representation of the fact that the field B_z increases along its own direction, so that there is a field gradient $\partial B_z/\partial z$. This gradient exerts on the magnetic moment a force $F = \mu\,\partial B_z/\partial z$ in the z-direction tending to slow down the translational motion in the helix. At a place where the work done by this force is equal to the original kinetic energy of translation, the longitudinal motion ceases, and only the circular motion remains. This occurs at a distance z_1 where $B = B_1$ such that

$$\tfrac{1}{2}mv_\parallel^2 = \int_0^{z_1} F\,dz = \int_0^{z_1} \mu(\partial B_z/\partial z)\,dz = \mu(B_1-B_0) \tag{7.47}$$

or from (7.42)

$$\left(\frac{v_\parallel}{v_\perp}\right)^2 = \frac{B_1-B_0}{B_0} \tag{7.48}$$

The angle $\alpha = \tan^{-1}(v_\perp/v_\parallel)$ between the total velocity of the particle and the direction of the magnetic field is called the pitch angle.

Equation (7.48) shows that if the pitch angle has a magnitude α_0 at $z = 0$, where $B = B_0$, the longitudinal motion in the helix will cease at a place where

$$B_1 = B_0/\sin^2\alpha_0 \tag{7.49}$$

and the particle will simply move in a circle. The force F then acts to restart the longitudinal motion in the reverse direction so that the particle retraces its path in a gradually opening helix. The place where the motion is reversed is called a *mirror point*. In the earth's magnetic

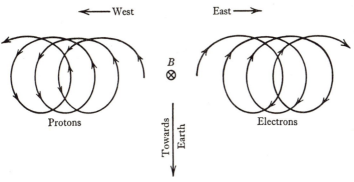

Fig. 7.4. Because the earth's magnetic field decreases upwards the path of a charged particle gyrating round it has a radius of curvature that is greater above than below so that the particle drifts sideways. The observer is supposed to be facing north so that the field is directed into the paper: protons drift to the west, electrons to the east.

field, where the lines of force converge towards both poles, there are mirror points at both ends, and particles can move repeatedly back and forth between them: they are said to be trapped in the geomagnetic field. Equation (7.49) shows that the position of the mirror point depends only on the distribution of the field and on the pitch angle of the helical motion: it is independent of the mass, the charge, or the energy of the particle.

Another motion of the guiding centre is produced by the radial decrease of the geomagnetic field: it can be understood with the help of fig. 7.4. Because the radius of curvature in the particle's circular motion is a little greater at points farther from the earth, where the field is weaker, the particle traverses a slightly looped path in the plane perpendicular to the magnetic field. The result is that the whole of the helical motion drifts bodily sideways round the earth. Since the direction of motion depends on the sign of the charge it is different for electrons and for positive ions. Detailed calculation shows that the drift velocity V_{d_1} is given by

$$V_{d_1} = \frac{mv_\perp^2}{2eBR} \qquad (7.50)$$

Here R is the radius of curvature of the field line; it comes into this expression because the radial gradient of B (which determines the drift discussed here) is related to its curvature.

Movement of the particle along a curved field line is also associated with another component of sideways drift in the following way. The velocity (V_{d_2}) of this drift, in a direction perpendicular to the field B, produces a force eBV_{d_2} that is along the radius of curvature of the field; this force must equal the centripetal force required to make the charge follow the curved field line. Hence

$$eBV_{d_2} = mv_\parallel^2/R$$

or
$$V_{d_2} = \frac{mv_\parallel^2}{eBR} \qquad (7.51)$$

The drift velocities V_{d_1} and V_{d_2} are in the same direction and add to give a total velocity

$$V_d = V_{d_1} + V_{d_2} = \frac{m}{eBR}(\tfrac{1}{2}v_\perp^2 + v_\parallel^2) \qquad (7.52)$$

It is determined by the kinetic energies $\tfrac{1}{2}mv_\perp^2$ and $\tfrac{1}{2}mv_\parallel^2$ of the particle appropriate to its velocities parallel and perpendicular to the field line, and for given energies in the non-relativistic range it is the same for protons and electrons. The oppositely directed drift motions of the protons and electrons produce a ring current, in which the current density is $2neV_d$, which from (7.52) is closely related to the total energy of the trapped particles.

It can be shown, by a quite general argument, that if energetic particles are confined to a ring surrounding the earth in the equatorial plane they carry a current proportional to their energy W, and this current produces a field (ΔB) on the equator at the earth's surface such that

$$\frac{\Delta B}{B_0} \doteqdot \frac{W}{W_{\text{ext}}} \qquad (7.53)$$

where B_0 is the undisturbed field at the earth's surface and W_{ext} is the energy of the geomagnetic field in the region outside the earth.

Suppose that the charged particles, with concentration n, are in a ring, with circular cross-section of area A, encircling the earth at a geocentric distance ρ and that, in virtue of their velocities (v) in the tangential direction they exert a pressure $p = nmv^2$. The force Ap, over the cross-section A, acting around the circumference of the ring, produces an outward force on unit length of the ring equal to Ap/ρ. This force is balanced by the force, Bi, on unit length of a current i flowing round the ring, so that

$$i = Ap/B\rho \qquad (7.54)$$

This ring current produces a field (ΔB) at its centre or, with sufficient accuracy, at the earth's surface, given by[†]

$$\Delta B/\mu_0 = i/2\rho = Ap/2\rho^2 B \qquad (7.55)$$

The tangential kinetic energy of the particles in unit volume of the ring is $\frac{1}{2}nmv^2 = \frac{1}{2}p$ and the volume of the ring is $2\pi\rho A$ so that the total kinetic energy is

$$W = \pi p\rho A \qquad (7.56)$$

Combination of (7.55) and (7.56) yields

$$\Delta B = \mu_0 W/2\pi\rho^3 B \qquad (7.57)$$

If B_0 is the unperturbed field at the earth's surface, radius R, then $\rho^3 B = R^3 B_0$ so that, from (7.57),

$$\Delta B/B_0 = \mu_0 W/2\pi R^3 B_0^2 \qquad (7.58)$$

Now it can also be shown easily that the total energy in the unperturbed field outside the earth is $W_{\text{ext}} = 4\pi B_0^2 R^3/3\mu_0$ so that (7.58) can be written

$$\Delta B/B_0 = 2W/3W_{\text{ext}} \qquad (7.59)$$

7.4 'Frozen-in' field lines [92]

An important theorem that applies to parts of the ionosphere where collisions are infrequent and conductivities are great is often called the

---

† μ_0 denotes the permeability of free space and must not be confused with the symbol μ that denotes the magnetic moment of a gyrating charged particle.

principle of *frozen-in field lines*. It is supposed that the field is represented by lines of induction drawn so that the number per unit area
passing through a small area perpendicular to the lines is equal to the
magnitude of the induction at that point. The principle can then be
stated as follows:

> When there are movements of the ionospheric plasma the magnetic
> field changes in such a way that the field lines move, and change
> their shape, as though they were carried with the plasma.

The theorem, which is widely used in discussion of the ionosphere, is
usually justified by a somewhat abstract mathematical argument. Some
students of the ionosphere who like to have a more physical picture may
find the following simple approach will help them to understand the
argument. In the mathematical treatment a start is usually made by
eliminating currents from the equations and concentrating on magnetic
fields and e.m.f.s. Here, in contrast, the emphasis is on the currents
that flow in the plasma.

Let us start by discussing an ordinary electrical circuit carrying a
current i that produces a magnetic field H, the relation between the
two quantities being given by $\oint H \, \mathrm{d}l = i$. Suppose there are no
batteries, so that the e.m.f. $\mathscr{E}$ round the circuit is given by $\mathscr{E} = -\mathrm{d}\phi/\mathrm{d}t$
where $\phi = \iint \mathbf{B} \cdot \mathrm{d}\mathbf{S}$ is the flux of induction through the circuit. This
flux may consist of a part ϕ_i produced by the current itself and a part
ϕ_x produced by external currents. If the inductance of the circuit is L
then by definition $\phi_i = Li$. If ϕ_x changes, the current changes as given
by

$$L \frac{\mathrm{d}i}{\mathrm{d}t} + Ri = -\frac{\mathrm{d}\phi_x}{\mathrm{d}t} \tag{7.60}$$

with a time constant L/R.

Consider a superconducting circuit in which R is vanishingly small,
and consider times short compared with the very long time constant,
so that (7.60) can be written

$$L \frac{\mathrm{d}i}{\mathrm{d}t} = -\frac{\mathrm{d}\phi_x}{\mathrm{d}t} \tag{7.61}$$

or $Li + \phi_x = $ constant and hence $\phi_i + \phi_x = \phi = $ constant. Thus, however ϕ_x may change, the current adjusts itself so that the total flux
through the circuit remains constant.

Notice that, if ϕ_x remains constant, the current persists, without any applied e.m.f., and produces a constant flux ϕ_i. If the circuit is moved, mechanically or electrically, the field is carried with it, unchanged; it is only when ϕ_x is changed that the current, and with it ϕ_i, changes.

Suppose next that the superconducting circuit can not only move but can also change its shape, and expand and contract, as though made of elastic. Its inductance might then change and it seems reasonable to suppose that (7.61) should then be written

$$\frac{d}{dt}(Li) = i\frac{dL}{dt} + L\frac{di}{dt} = -\frac{d\phi_x}{dt} \qquad (7.62)$$

and if we can write $\phi_i = Li$, as before, then

$$\frac{d\phi_i}{dt} = -\frac{d\phi_x}{dt} \qquad (7.63)$$

If this supposition were correct then (7.63) shows that again the current alters so as to keep the total flux through itself constant.

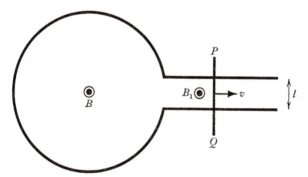

Fig. 7.5. To illustrate the phenomena in a circuit that can change its shape.

The justification for the important argument summarized by (7.62) and (7.63) is usually presented in a generalized vector form. Let us understand it in a simpler way by considering a circuit that can change its shape because it has a part consisting of a wire PQ that can slide over stationary parallel wires a distance l apart, as in fig. 7.5. Suppose also that there is magnetic induction B (not necessarily uniform) perpendicular to the plane of the circuit, and that it has the magnitude B_1 near PQ. Let PQ move outwards with velocity v so that a field $E = vB_1$ is induced in it. There is then an e.m.f. $\mathscr{E} = (vB_1)l$ which causes the current in the circuit to change at a rate given by $L\,di/dt = \mathscr{E}$. The change of current is in a sense to reduce the flux ϕ_i ($= Li$) through the circuit at a rate $-d\phi_i/dt = \mathscr{E} = (vB_1)l$. But the area of the circuit is also changing at a rate vl so that flux is being added at a rate $d\phi_i/dt = B_1(vl)$. This is equal and opposite to the rate of reduction resulting from the induced e.m.f., and the total flux remains unaltered.

If there is no change in the externally produced flux (ϕ_x), but the shape of the circuit changes, so that the inductance changes from L_1 to L_2, the current changes from i_1 to i_2 where $L_1 i_1 = L_2 i_2$. This change is caused by the e.m.f.s induced when one part of the circuit moves through the field produced by the currents in other parts.

Now let the foregoing discussion apply to a thin ring of super-conducting fluid that forms part of an extended volume of super-conductor in which persistent currents are flowing. Currents may flow around and across the ring, but only those flowing around it produce the flux ϕ_i: currents in the rest of the material produce the flux ϕ_x. Now let the ring move and change its shape so that ϕ_i changes, and let the rest of the material move so that ϕ_x changes. The foregoing discussion then shows that the total flux ϕ through the ring remains constant. If, then, lines of induction are drawn so that the number passing through the ring represents the magnitude of the flux through it, the number remains constant however the ring moves or changes its shape. If this statement is extended to include all possible rings drawn in the fluid it follows that the lines of induction are constrained to move with the fluid.

It should be noticed that if there is resistance in the circuit, a term Ri must be inserted on the left-hand side of (7.61) and (7.62) and in the discussion that follows (7.63), $\mathscr{E}$ must be equated to $Ri + B_1 vl$: it then no longer follows that the flux remains constant. The freezing-in theorem is thus valid only for fluids with very small resistance.

8 Electromagnetic, hydromagnetic and electro-acoustic waves†

8.1 Introduction

Waves in which electric and magnetic fields, ions and electrons all take part are important in the ionosphere partly because nature uses them to influence the behaviour of that 'sphere' and partly because man uses them as experimental tools in his investigations. The student of the ionosphere does not need to know about all the complications that can arise when waves of different frequency travel in different directions relative to the earth's magnetic field: indeed many important phenomena can be understood by considering the motions of electrons and ions in some simple situations where waves travel only along, or perpendicular to, the geomagnetic field. This chapter discusses these motions to show in terms of simple physical concepts how the following wave phenomena arise:

1. Refractive index less than unity, leading to reflection of waves.
2. The limitation of wave propagation to certain frequency ranges (pass-bands).
3. Refractive index infinite or zero at ends of the pass-bands.
4. Electron- and ion-acoustic waves similar in many ways to sound waves in a neutral gas.

The presence of the geomagnetic field causes the ionosphere to be anisotropic so that when an electro-magnetic wave traverses it the energy, or a small wave-packet, travels in a direction that is different, in general, from the wave-normal direction; it is called the ray direction. The relation between the wave-normal direction and the ray direction is discussed in §8.8.

The man-made waves that are important for the present purposes have frequencies in the range from about 10 kHz to 50 MHz; they are used for exploring the ionosphere from the ground, or from space vehicles, or for radio communication. Because the efficient radiation

† 3, 8, 14, 15, 17, 26, 30.

[135]

of power requires a radiator with dimensions of the order of one wave-length, man-made senders do not use frequencies much less than 10 kHz (wavelengths greater than 30 km). Nature, however, can make use of mechanisms that extend over great distances and can radiate waves of correspondingly great length. Their frequencies cover a wide range from a few millihertz to a few megahertz. Those of smallest frequency originate in the vast spaces of the magnetosphere, others, of greater frequency, originate in earth-bound lightning flashes.

As waves pass through the ionosphere the electrons and ions partake of their motion so that energy is shared between the fields of the wave and the kinetic energy of the charged particles. At the greater frequencies it resides mainly in the electric and magnetic fields and the wave is called electromagnetic: at the smaller frequencies it is mainly in the kinetic energy of the particles and in the magnetic field; the wave is then called hydromagnetic.

8.2   Properties common to all electric waves

Maxwell's equations can be used to show that electromagnetic and hydromagnetic waves of all frequencies travelling through an ionized medium in the presence of a steady magnetic induction (B) have some fundamental properties in common. Suppose a plane wave has its wave normal in the positive z-direction and has angular frequency ω and angular wave number k so that all quantities vary like $\exp i(\omega t - kz)$ and $d/dx = d/dy = 0$. The wave, with electric field **E** and magnetic field **H**, moves the charges in the medium so that a volume polarization† **P** is produced and the electric displacement is $\mathbf{D} = \epsilon_0 \mathbf{E} + \mathbf{P}$. The medium may be electrically anisotropic so that the polarization is not necessarily in the direction of the applied field. Maxwell's equations can then be written as follows

$$\operatorname{curl} \mathbf{H} = \dot{\mathbf{D}} = \epsilon_0 \dot{\mathbf{E}} + \dot{\mathbf{P}} \tag{8.1}$$

† The word polarization is used with two different meanings: possible confusion is avoided by writing: *volume polarization* (P) to mean the electric dipole moment per unit volume, and *wave polarization* to mean the pattern traced out by one of the oscillating components of the wave-field.

$$\left\{ \begin{array}{ll} ikH_y = i\omega(\epsilon_0 E_x + P_x) & (8.2) \\ -ikH_x = i\omega(\epsilon_0 E_y + P_y) & (8.3) \\ o = i\omega(\epsilon_0 E_z + P_z) & (8.4) \end{array} \right.$$

or

$$\mathrm{curl}\,\mathbf{E} = -\mu_0 \dot{\mathbf{H}} \tag{8.5}$$

or

$$\left\{ \begin{array}{ll} ikE_y = -i\omega\mu_0 H_x & (8.6) \\ -ikE_x = -i\omega\mu_0 H_y & (8.7) \\ o = -i\omega\mu_0 H_z & (8.8) \end{array} \right.$$

The following quite general conclusions can be drawn from these equations.

1. From (8.8) $H_z = 0$ so that the magnetic field of the wave lies entirely in the wave front.

2. From (8.4) $P_z = -\epsilon_0 E_z$ so that there may be a component of the electric field normal to the wave front: it is accompanied by a component of volume polarization (P) in the same direction and of opposite sense, that balances it so as to make the displacement (D) lie entirely in the wave front.

3. The ratio H_y/H_x represents the nature of the wave polarization; if it is real the wave is plane polarized, if it is complex the polarization is elliptical. From (8.6) and (8.7) we find $E_x/E_y = -H_y/H_x$ so that the magnetic field and the component of the electric field that lies in the wave front are always at right angles to each other, and trace out similar ellipses, with their major axes perpendicular to each other.

4. Division of (8.2) by (8.3) and insertion of the relation

$$H_y/H_x = -E_x/E_y$$

leads to

$$P_x/P_y = E_x/E_y \tag{8.9}$$

5. From (8.7) $H_y = (k/\mu_0 \omega) E_x$ and insertion of this into (8.2) yields $P_x/\epsilon_0 E_x = (k^2/\mu_0 \epsilon_0 \omega^2) - 1$. Now $\omega/k = V$, the velocity of the wave, and $(\mu_0 \epsilon_0)^{-\frac{1}{2}} = c$, the velocity of electromagnetic waves in free space, so that $k^2/\omega^2\mu_0\epsilon_0 = c^2/V^2 = n^2$ where n is the refractive index of the medium. Hence (if (8.9) is also used)

$$n^2 - 1 = P_x/\epsilon_0 E_x = P_y/\epsilon_0 E_y \tag{8.10}$$

To find the refractive index it is hence sufficient to calculate P_x/E_x or P_y/E_y for any particular situation.

6. From (8.2)

$$H_y = (\omega\epsilon_0/k) \quad (1 + P_x/\epsilon_0 E_x)\, E_x$$

$$= n^{-1}(\epsilon_0/\mu_0)^{\frac{1}{2}} \quad n^2 \quad E_x$$

$$= n(\epsilon_0/\mu_0)^{\frac{1}{2}} E_x$$

and similarly from (8.3)   (8.11)

$$H_x = -n(\epsilon_0/\mu_0)^{\frac{1}{2}} E_y$$

The total magnetic field $\sqrt{(|H_x|^2 + |H_y|^2)}$ is thus equal to $n(\epsilon_0/\mu_0)^{\frac{1}{2}}$ times the total electric field $\sqrt{(|E_x|^2 + |E_y|^2)}$ in the wave front.

It is to be noted that, whereas the magnetic field lies entirely in the wave front, the electric field can have a component along the wave normal. This difference arises because the medium is supposed to be electrically anisotropic so that $\mathbf{P}$ is not necessarily in the direction of the electric field $\mathbf{E}$. If, in addition, it were supposed to be magnetically anisotropic, so that the magnetic polarization ($\mathbf{I}$) was not necessarily in the direction of the magnetic field ($\mathbf{H}$), there could be a component of $\mathbf{H}$ along the wave normal. In all cases, however, the electric displacement $\mathbf{D} = \epsilon_0 \mathbf{E} + \mathbf{P}$ and the magnetic induction $\mathbf{B} = \mu_0 \mathbf{H} + \mathbf{I}$ lie in the wave front.

8.3   Conditions for zeros and infinites in the refractive index

If the volume polarization P is in phase (or antiphase) with E, so that P/E is a real quantity, (8.10) shows that n^2 is real and the wave suffers no absorption as it travels. In practice, however, collisions between electrons and heavy particles produce a situation in which P and E are neither in phase nor in antiphase, so that the ratio P/E, and n, are complex. There is then attenuation of the wave and if n is written $n = \mu - i\chi$, μ is the (real) refractive index and χ is the attenuation constant. The conditions that would make $n = 0$ in the absence of collisions, or μ very small in the presence of collisions, are important because they

(*a*) result in reflection of a wave travelling vertically into a horizontally stratified ionosphere,

(b) may correspond to the limiting frequency at the end of a pass-band.

For the present purpose it is sufficient to ignore collisions and to examine the conditions that make $n = 0$.

The conditions that make n infinite (or μ very great, if there are collisions) are also important because they mark the end of a pass-band of frequencies. They also often correspond to a small group velocity and thus represent a situation where a wave pulse emitted by a satellite can travel with the satellite for a measurable time: from observations made in that situation it is often possible to make useful deductions about the surrounding plasma (§9.2.1).

This section examines the conditions that make the refractive index (n) zero, and those that make it infinite. Equation (8.10) shows that the refractive index is equal to zero when

$$P_x = -\epsilon_0 E_x \quad (\text{or } P_y = -\epsilon_0 E_y)$$

Much use will be made of this condition in what follows.

The refractive index becomes very great when $P_x \gg \epsilon_0 E_x$ (and $P_y \gg \epsilon_0 E_y$): under these conditions E_x and E_y become very small, and the only appreciable electric field is E_z perpendicular to the wave front. The anisotropy in the medium is then such that this field, directed along the wave normal, moves the charges in the plane of the wave front so that P_x (and/or P_y) is finite although E_x and E_y are zero.

When the refractive index is very small (8.11) shows that H/E is much less than its value $(\epsilon_0/\mu_0)^{\frac{1}{2}}$ in free space, so that the electric field predominates. This is because the wavelength is then very great and $H \propto \text{curl } E = dE/dz$ is very small (see (8.5)).

When the refractive index is very large (8.11) shows that

$$H_y/E_x \gg (\epsilon_0/\mu_0)^{\frac{1}{2}}, \quad \text{and} \quad H_x/E_y \gg (\epsilon_0\mu_0)^{\frac{1}{2}}$$

indeed since E_x and E_y are very small this result might be expected: it might also be expected because the wavelength is very small so that dE/dz ($= \text{curl } E \propto \mu_0 H$) is great. The smallness of the transverse electric fields E_x and E_y does not necessarily imply that the longitudinal field E_z is small.

8.4   The motions of the electrons

Up to this stage the discussion has been quite general; it has been
supposed only that the wave produces a volume electric polarization
of some kind in the medium. We now turn to discuss the nature of this
polarization when the medium consists of electrons of mass m and
charge e moving freely (without collisions) amongst a neutralizing
assembly of positive ions, with the earth's steady magnetic field superimposed: we describe the field in terms of its induction B. At first we

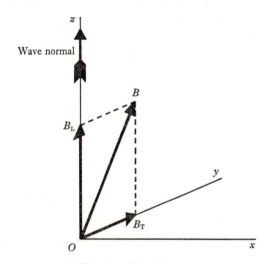

Fig. 8.1. Co-ordinate axes.

suppose that the positive ions are fixed in position and do not move
as the wave advances: it will be shown later that this is effectively what
happens at frequencies greater than about 10 kHz that are usually
considered to be part of the radio spectrum. Later this restriction is
relaxed to permit the discussion of waves of smaller frequency that are
usually considered to be part of the spectrum of hydromagnetic
waves.

Suppose that B is in the zy plane with components B_L (longitudinal)
along Oz and B_T (transverse) along Oy, and that the wave normal is
along the positive z-direction as in fig. 8.1. The equations of motion of

an electron, usually called the constitutive relations of the medium, are then as follows:

$$m\ddot{x} = eE_x + eB_L\dot{y} - eB_T\dot{z} \qquad (8.12\,a)$$

$$m\ddot{y} = eE_y - eB_L\dot{x} \qquad (8.13\,a)$$

$$m\ddot{z} = eE_z + eB_T\dot{x} \qquad (8.14\,a)$$

For an electron e is, of course, negative.

Now suppose that the fields and the displacements of the electrons vary simple-harmonically as represented by $e^{i\omega t}$, and that there are N electrons per unit volume.† In what follows it will be necessary to evaluate the components of the volume polarization

$$P_{x(y.z)} = Nex,(y,z),$$

so that (8.12 a) (8.13 a) and (8.14 a) can be usefully written

$$-\omega^2 P_x = \frac{Ne^2}{m}E_x + i\omega\frac{eB_L}{m}P_y - i\omega\frac{eB_T}{m}P_z \qquad (8.12\,b)$$

$$-\omega^2 P_y = \frac{Ne^2}{m}E_y - i\omega\frac{eB_L}{m}P_x \qquad (8.13\,b)$$

$$-\omega^2 P_z = \frac{Ne^2}{m}E_z + i\omega\frac{eB_T}{m}P_x \qquad (8.14\,b)$$

It is usual to write these equations in terms of parameters defined as follows.

$\Omega = |e|B/m =$ the gyro-frequency with which an electron or ion rotates freely round the imposed magnetic field.

$Y = \Omega/\omega \qquad Y_L$ and Y_T are the values of Y obtained by writing B_L and B_T in place of B.

$\omega_0 = (Ne^2/\epsilon_0 m)^{\frac{1}{2}}$

$X = \omega_0^2/\omega^2$

These parameters are independent of the sign of e, so that when they are substituted into (8.12 b), (8.13 b) and (8.14 b) the equations take different forms for electrons and for positive ions. Since the parameters

† In this chapter the electron concentration is denoted by N, and not by n as elsewhere in the book, to avoid confusion with the symbol n used for the refractive index.

have different magnitudes for electrons and ions they are labelled with the appropriate subscripts (e and i). The equations can then be written

For electrons

$$\epsilon_0 X_e E_x = -P_x + iY_{eL}P_y - iY_{eT}P_z \qquad (8.12c)$$

$$\epsilon_0 X_e E_y = -P_y - iY_{eL}P_x \qquad (8.13c)$$

$$\epsilon_0 X_e E_z = -P_z + iY_{eT}P_x \qquad (8.14c)$$

For positive ions

$$\epsilon_0 X_i E_x = -P_x - iY_{iL}P_y + iY_{iT}P_z \qquad (8.12d)$$

$$\epsilon_0 X_i E_y = -P_y + iY_{iL}P_x \qquad (8.13d)$$

$$\epsilon_0 X_i E_z = -P_z - iY_{iT}P_x \qquad (8.14d)$$

In what follows deductions are made chiefly from the equations for electrons; in order to discuss positive ions it is necessary simply to alter the sign of Y in the resulting expressions.

First consider a wave that travels through a medium where only electrons can move, and where there is no applied field ($B = 0$, $Y = 0$). (8.12c) and (8.13c) then show that $P_x/\epsilon_0 E_x = P_y/\epsilon_0 E_y = -X_e$ so that a wave with components of E and P in the wave front will travel with a refractive index given by

$$n^2 = 1 - X_e \qquad (8.15)$$

(8.14c) becomes $\epsilon_0 X_e E_z = -P_z$, but the general condition (8.4) requires $\epsilon_0 E_z = -P_z$: it follows that either $P_z = E_z = 0$, or $X_e = 1$. The solution $E_z = P_z = 0$ combined with the previous result concerning $E_{x,y}$ and $P_{x,y}$ shows that the wave represented by (8.15) is purely transverse. But if $X_e = 1$ there can also be a field component E_z perpendicular to the wave front: (8.15) then shows that the refractive index is zero, the velocity is infinite and the phases of all wave components are the same everywhere. The medium oscillates as a whole at a frequency, $\omega = \omega_{0e}$, that makes $X_e = 1$: it is called the *plasma frequency*.

The bodily oscillation of the plasma can be understood by supposing that all the electrons in a finite rectangular block are displaced a distance x normal to one pair of faces. Surface densities of charge equal to $\pm Nex$ then appear on these faces and produce a field $E = -Nex/\epsilon_0$ tending to restore the electrons in the body of the block to their original positions. The equation of motion of one of these electrons is thus $Ee = m\ddot{x}$, or $-Ne^2x/\epsilon_0 = m\ddot{x}$: it therefore oscillates simple-harmonically with frequency

$$\omega_{0e} = (Ne^2/\epsilon_0 m)^{\frac{1}{2}} \qquad (8.16)$$

corresponding to $X_e = 1$.

In a plasma in which electrons and ions are in thermal motion with gas kinetic velocities (v) the distance (l) travelled, during one plasma oscillation, by an electron moving with the gas-kinetic velocity is given by $l = 2\pi v/\omega_{0e}$. With $v = (2kT/m)^{\frac{1}{2}}$ and ω_{0e} from (8.16),

$$l = 2^{\frac{3}{2}}\pi(\epsilon_0 kT/Ne^2)^{\frac{1}{2}}$$

The length $\qquad \lambda_D = (\epsilon_0 kT/Ne^2)^{\frac{1}{2}} \qquad (8.17)$

which is a simple multiple of l is called the *Debye length*: since it is independent of the particle's mass, it is the same for ions and electrons (see also p. 213).

The wave that has E and P entirely in the wave front has refractive index given by (8.15) which can be written

$$n^2 = 1 - Ne^2/\epsilon_0 m\omega^2 = 1 - \omega_{0e}^2/\omega^2 = 1 - X_e \qquad (8.18)$$

The following points are of interest:

(a) The extent to which the square of the refractive index departs from unity is inversely proportional to the mass m of the charged particles. Thus when electrons and heavy ions are present in equal numbers the effect of the electrons far outweighs that of the ions: it is for that reason that the ions were supposed to be fixed in position in the above calculations.

(b) The refractive index is less than unity: this is because P/E is negative. To understand how this comes about it is convenient to discuss the situation first in terms of a positive charge for which the force, and thus the acceleration, is in the direction of E. In the simple harmonic motion of the wave under discussion the displacement is always in a direction opposite to the acceleration, so that the displacement x is in the direction opposite to E and the ratio $P/E (= Nex/E)$ is negative, and the refractive index is less than unity. If the charge were negative the force (and acceleration) would be in a direction opposite to E, the displacement would be in the same sense as E, but the polarization $(P = Nex)$ which depends on the sign of e, would be in the sense opposite to E and once again P/E would be negative.

(c) The mean energy densities W_E in the electric field and W_H in the magnetic field are given by

$$W_E = \tfrac{1}{2}\epsilon_0 E^2 \qquad (8.19)$$
$$W_H = \tfrac{1}{2}\mu_0 H^2,$$

and substitution from (8.11) gives

$$W_H = \tfrac{1}{2}\mu_0 n^2(\epsilon_0/\mu_0) E^2 = n^2 W_E \qquad (8.20)$$

where the appropriate symbols represent r.m.s. values. The mean density W_K of the kinetic energy of the charged particles is

$$W_K = N\tfrac{1}{2}mv^2 = \tfrac{1}{2}Nm\omega^2 x^2$$
$$= \frac{1}{2\epsilon_0}\frac{P^2}{X_e} \qquad (8.21)$$

Thus
$$\frac{W_K}{W_E} = \frac{P^2}{\epsilon_0^2 E^2 X_e} = |1 - n^2| \qquad (8.22)$$

from (8.10) and (8.18). Equations (8.20), (8.21) and (8.22) show that

(i) when $n \fallingdotseq 1$, $W_H \fallingdotseq W_E \gg W_K$: the energy resides equally in the electric and the magnetic fields and the wave is called 'electromagnetic';

(ii) when $n \fallingdotseq 0$, $W_H \ll W_E$ and $W_K \fallingdotseq W_E$: the energy resides equally in the particles' motion and in the electric field, the wave might be called 'electrokinetic'.

8.5   Longitudinal propagation in a magnetic field†

8.5.1   Only electrons moving

Next consider the effect of superimposing a magnetic field along the z-direction of the wave normal by writing $Y_e = Y_L$ and $Y_T = 0$ in

† The adjectives 'longitudinal' and 'transverse' have caused confusion. Ionospheric workers usually speak of longitudinal or transverse waves when they mean waves travelling along, or perpendicular to, a magnetic field, whereas plasma theorists use the adjectives to describe waves in which the electric field is along, or perpendicular to, the wave normal. Here the ionospherists' nomenclature is used and the expressions 'longitudinal propagation' or 'transverse propagation' imply that the wave normal is along, or perpendicular to, the imposed magnetic field.

(8.12c), (8.13c), (8.14c), so that

$$\epsilon_0 X_e E_x = -P_x + iY_e P_y \qquad (8.23)$$

$$\epsilon_0 X_e E_y = -P_y - iY_e P_x \qquad (8.24)$$

$$\epsilon_0 X_e E_z = -P_z \qquad (8.25)$$

To solve these equations we require other relations between the variables. Let us suppose that all perform the same wave motion represented by $\exp\{i(\omega t - kz)\}$ so that quantities that depend on the ratios of the variables will not change as the wave travels: non-changing quantities of that kind are exemplified by the state of polarization of the electric and magnetic fields and the relations of these fields to each other and to the volume polarization of the medium. With these restrictions we shall show that two waves, with different velocities and wave polarizations can travel with the same frequency and in the same direction: they are called *characteristic waves*. Each characteristic wave maintains its own proper polarization as it travels, but if a wave with the same frequency and wave-normal direction has a polarization different from one of the characteristic ones its polarization changes as it travels. A wave of that kind can be discussed by considering it to be composed of the two characteristic waves each of which travels with its own velocity.

When all quantities vary like $\exp\{i(\omega t - kz)\}$ it was shown in §8.2 that $\epsilon_0 E_z = -P_z$ and $P_x/E_x = P_y/E_y$. The condition $\epsilon_0 E_z = -P_z$ can be satisfied simultaneously with (8.25) only if $X_e = 1$ or if $E_z = P_z = 0$. The situation when $X_e = 1$ is discussed in §8.4: attention is here restricted to waves in which $E_z = P_z = 0$ so that the electric and magnetic fields and the polarization are all in the wave front.

Suppose that $P_x/P_y = a = E_x/E_y$ and substitute into (8.23) and (8.24) to show that $a = \mp i$.[†] Hence if all the variables are to have the same wave motion it must be possible to represent the electric field (E) and the polarization (P) as vectors of constant length rotating in one of the two possible senses; thus in a characteristic wave both these quantities are circularly polarized with right- or left-handed sense of rotation.

† The $\mp$ signs are represented in this order so that the equations that follow will take their usual form.

To find the refractive indices of the two possible waves insert the ratios $P_x/P_y = \mp i$ into (8.23) and (8.24) to show that

$$\frac{P_x}{\epsilon_0 E_x} = \frac{P_y}{\epsilon_0 E_y} = \frac{-X_e}{1 \pm Y_e} \tag{8.26}$$

Hence

$$n^2 = 1 + \frac{P_x}{\epsilon_0 E_x} = 1 - \frac{X_e}{1 \pm Y_e} \tag{8.27}$$

This expression can be rewritten

$$n^2 = 1 - \frac{\omega_{0e}^2}{\omega(\omega \pm \Omega_e)} \tag{8.28}$$

from which it is seen that $n^2 = 0$ for one or other of the waves when $\omega(\omega \pm \Omega_e) = \omega_{0e}^2$
or when

$$\left. \begin{array}{l} \omega = \omega_{1e} = (\omega_{0e}^2 + \tfrac{1}{4}\Omega_e^2)^{\frac{1}{2}} - \tfrac{1}{2}\Omega_e \\[4pt] \omega = \omega_{2e} = (\omega_{0e}^2 + \tfrac{1}{4}\Omega_e^2)^{\frac{1}{2}} + \tfrac{1}{2}\Omega_e \end{array} \right\} \tag{8.29}$$

or

It is also seen that $n^2 = \infty$ for the wave corresponding to the lower sign when $\omega = \Omega_e$ and that $n^2 = \pm\infty$ when $\omega = 0$.

It is convenient sometimes to show how n^2 depends on the plasma frequency when the wave frequency remains constant and sometimes to show how it depends on the wave frequency when the plasma frequency remains constant. Fig. 8.2 shows n^2 plotted against $\omega_{0e}^2/\omega^2 (= X_e)$ and can be used to consider the way in which the refractive index varies with the electron concentration $N (\propto \omega_{0e}^2)$ when the wave frequency and the gyro-frequency remain constant. It is useful, for example, in discussing how a wave of fixed frequency behaves as it travels upwards into the ionosphere where the electron concentration depends on the height. If $\omega > \Omega_e (Y_e < 1)$ the refractive indices of both waves are less than unity, whereas if $\omega < \Omega_e (Y_e > 1)$ the refractive index of one is greater than unity.

To illustrate the form of dispersion curves, that relate n^2 to ω, it is useful to assign simple numbers, that have no relevance to the ionosphere, to the different frequencies. To illustrate the relation (8.28) between n^2 and ω fig. 8.3 (a) is drawn for a situation where $\omega_{0e} = 10$, $\Omega_e = 4$. The waves corresponding to each sign in the equation can travel at frequencies greater than their respective cut-off frequencies ω_{1e} and ω_{2e}: in these high frequency pass-bands the refractive indices are less than unity. The wave corresponding to the lower sign can also travel in a low frequency pass-band, extending from Ω_e down to zero, in this pass-band the refractive index is greater than unity, and at the two limiting frequencies it is infinite. This wave, with $n > 1$, is often called *the whistler mode*, it is discussed below.

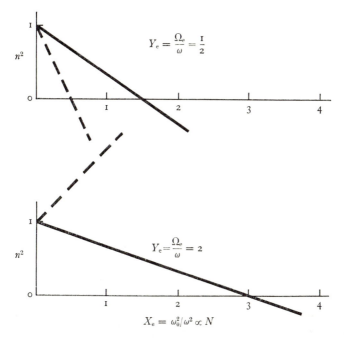

Fig. 8.2. The square of the refractive indices plotted against X_e (proportional to electron concentration) for the two characteristic waves with wave normals along the direction of the imposed magnetic field as given by (8.27). The continuous lines correspond to the upper sign. The dashed lines, corresponding to the lower sign, take different forms according as $\Omega_e < \omega$ or $\Omega_e > \omega$ ($Y_e < $ or > 1).

8.5.2 Refractive index greater than unity (the whistler mode)

To examine the conditions that make the refractive index greater than unity and sometimes infinite, consider fig. 8.4 which represents the motion of a *positively* charged particle acted upon by the electric field E of the wave, constant in magnitude and rotating in one sense or the other with angular frequency ω around the imposed magnetic field.

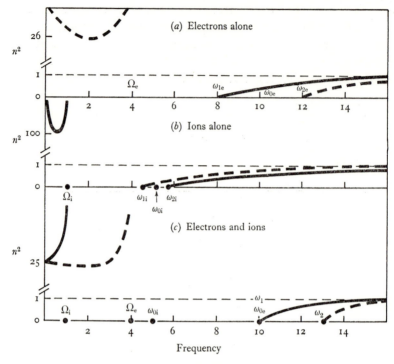

Fig. 8.3. Dispersion curves (n^2 as a function of ω) for longitudinal propagation drawn for values of the parameters that are very different from the ionospheric ones. The continuous and dashed curves correspond, respectively, to the upper and lower signs in (8.34 a). The three sets of curves correspond to situations where (a) only electrons can move ($\omega_{0e} = 10$, $\Omega_e = 4$, $\omega_{0i} = \Omega_i = 0$) – equation (8.28); (b) only ions can move ($\omega_{0i} = 5$, $\Omega_i = 1$, $\omega_{0e} = \Omega_e = 0$); (c) both electrons and ions can move ($\omega_{0e} = 10$, $\omega_{0i} = 5$, $\Omega_e = 4$, $\Omega_i = 1$, $m_e/m_i = 4$). The curve in (c) is obtained by adding $(1 - n^2)$ for the curves in (a) and (b).

The charge moves with the same angular frequency in a circle of radius r, the force $e(E \pm B\omega r)$ providing the necessary acceleration towards the centre so that

$$eE \pm Be\omega r = m\omega^2 r \qquad (8.30)$$

The ratio between the force $Be\omega r$ and the required centripetal force $m\omega^2 r$ is $Be/m\omega = \Omega_e/\omega \, (= Y_e)$: thus if $\omega = \Omega_e$, and if the direction of motion corresponds to the upper sign in (8.30), the force of the

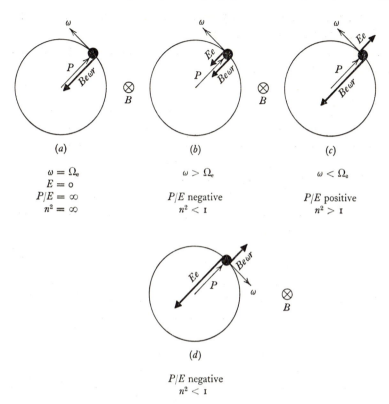

Fig. 8.4. To illustrate the forces on a positive charge acted upon by an electric field of constant strength E rotating with angular velocity ω in a plane perpendicular to a magnetic induction B directed into the paper.

magnetic field alone keeps the charge moving in the circle (fig. 8.4(a)). Under these conditions (8.30) is satisfied with $E = 0$ so that $P/E = \infty$ and $n^2 = \infty$. The direction of rotation considered here is appropriate to the characteristic wave represented by the dashed line in fig. 8.3(a). If, however, $\omega > \Omega_e$ the force $Be\omega r$ is too small to provide the centripetal acceleration and an additional force eE is required, also directed to the centre, as shown in fig. 8.4(b): the direction of E is then opposite to the direction of P (proportional to the displacement of the charge from the centre) so that P/E is negative and $n^2 < 1$. But if $\omega < \Omega_e$ the force $Be\omega r$ is too great (fig. 8.4(c)), and a force eE is required, directed away from the centre, to satisfy (8.30): E is then in the same direction as the charge's displacement P, and $n^2 > 1$.

If the field E, and the charge, rotate in the sense opposite to that considered above (8.30) becomes

$$eE - Be\omega r = m\omega^2 r \qquad (8.31)$$

and the appropriate characteristic wave is that of the continuous line in fig. 8.3 (*a*). The force $Be\omega r$ is then radially outwards (fig. 8.4(*d*)) and an inwardly directed field E is required at all frequencies to satisfy (8.31): thus P/E is negative and $n^2 < 1$ for all frequencies.

If the frequency is very small the equation of motion becomes $eE = \mp Be\omega r$ and the radius $r = \mp E/B\omega$ is very great, as also is $P = ner$. Hence P/E, and n^2, are very great and positive or negative according to the sign in the equation.

Fig. 8.4 shows that in the wave for which $n^2 > 1$ at some frequencies [(*a*) and (*c*)] the direction of rotation of the field vectors is related to the direction of the imposed field as the rotation is related to the translation of a left-handed screw. If the wave travels along the positive direction of the field there is the same relation between the rotation of the field vectors and the direction of travel of the wave and it is said to be circularly polarized with a left-handed sense.

In the ionosphere, where the operative electrons have negative charge, the diagrams of fig. 8.4 still apply if the field B is reversed. It follows that in the ionosphere, the right-handed circularly polarized wave, travelling along the positive field direction, is the one that has refractive index greater than unity for some frequencies.

8.5.3. Electrons and ions both moving. Hydromagnetic waves [83, 103]

Up to this stage it has been supposed that the positive ions are fixed in position and only the electrons move: suppose now that the positive ions are also free to move. Their equations of motion are similar to those of the electrons. It is important to notice that the movement of charges of one sign does not build up any space charge capable of producing an electric field that could act on charges of the other sign.†

† Although in the situation discussed here the ions and electrons do not influence each other's motions, the situation is quite different if there is a component of the imposed magnetic field in the wave front (see §8.6.1).

The equations of motion of the ions and of the electrons differ only because (*a*) the signs of the charges are different, and (*b*) the masses and hence the gyro-frequencies are different. It is thus convenient to use a similar nomenclature for ions and electrons, denoting the two types of particles by the subscripts i and e. When the equations are written in the form of (8.23), (8.24) and (8.25), it is necessary to reverse the sign of Y to represent an ion with a positive charge. A simple extension of the previous analysis (8.26) and (8.27), then shows that

$$P_x(\text{total}) = P_{xe} + P_{xi} = -\left(\frac{X_e}{1 \pm Y_e} + \frac{X_i}{1 \mp Y_i}\right)\epsilon_0 E_x$$

and

$$n^2 - 1 = -\left(\frac{X_e}{1 \pm Y_e} + \frac{X_i}{1 \mp Y_i}\right) \tag{8.32}$$

The same result is obtained for the y-components. The curve that represents $(n^2 - 1)$ is thus the sum of the two curves that would represent this quantity if first one, and then the other, type of particle were alone capable of moving (see fig. 8.3). It should be noticed that, in combining the curves for electrons and ions it is the departures of n^2 from unity that must be added, and not the values of n^2 itself.

If use is made of the relation $X_e Y_i = X_i Y_e$ and if we write

$$X = X_e + X_i \tag{8.33}$$

or the equivalent

$$\omega_0^2 = \omega_{0e}^2 + \omega_{0i}^2 \tag{8.33a}$$

then (8.32) can be written

$$n^2 = 1 - \frac{X}{(1 \pm Y_e)(1 \mp Y_i)} \tag{8.34}$$

or

$$n^2 = 1 - \frac{\omega_0^2}{(\omega \pm \Omega_e)(\omega \mp \Omega_i)} \tag{8.34a}$$

From (8.34a) it is seen that the frequencies ω_1 and ω_2 that make $n = 0$ are given by

$$\omega_1 = \{\omega_0^2 + \tfrac{1}{4}(\Omega_e + \Omega_i)^2\}^{\frac{1}{2}} - \tfrac{1}{2}\{\Omega_e - \Omega_i\} \tag{8.35}$$

and

$$\omega_2 = \{\omega_0^2 + \tfrac{1}{4}(\Omega_e + \Omega_i)^2\}^{\frac{1}{2}} + \tfrac{1}{2}\{\Omega_e - \Omega_i\} \tag{8.36}$$

Their magnitudes when only electrons can move, given by writing $\Omega_i = 0$, are seen to be the same as those of (8.29).

The dispersion curves for electrons alone, ions alone, and electrons and ions together, are illustrated in fig. 8.3 for an example in which

6

$m_i/m_e = 4$, $\omega_{0e} = 10$, $\omega_{0i} = 5$, $\Omega_e = 4$, $\Omega_i = 1$. The curve for electrons and ions together is obtained by adding $n^2 - 1$ on the other curves. The shape is obvious except for the behaviour as the frequency tends to zero, but then curves that go to infinity, one on the positive and the other on the negative side, have to be added and the result needs further consideration as follows. Equation (8.34a) shows that, when $\omega \to 0$

$$n^2 - 1 = \omega_0^2/\Omega_e\Omega_i \tag{8.37}$$

for both signs, so that both waves have the same refractive index. Since $\omega_0^2 \doteqdot \omega_{0e}^2$ in the ionosphere, and $\omega_{0e}^2 = \omega_{0i}^2\Omega_e/\Omega_i$, (8.37) can be written

$$n^2 - 1 \doteqdot \omega_{0i}^2/\Omega_i^2$$

or, since $\omega_{0i}^2/\Omega_i^2 \gg 1$, $n \doteqdot \omega_{0i}/\Omega_i$ (8.38)

This refractive index is appropriate to both characteristic waves when the frequency is small enough to make $\omega^2 \ll \Omega_i^2$. A wave of any polarization, made up by some combination of the two, thus travels with its form unchanged.

To consider the distribution of energy in this low-frequency wave for which $n^2 \gg 1$, substitute the expressions for P_e and P_i from (8.32) into (8.21) to obtain

$$W_{Ke} = \frac{1}{2\epsilon_0}\left(\frac{X_e^2}{Y_e^2}\right)\frac{1}{X_e}\epsilon_0 E^2$$

$$= (\omega_{0e}^2/\Omega_e^2)W_E \tag{8.39}$$

and $$W_{Ki} = (\omega_{0i}^2/\Omega_i^2)W_E \tag{8.40}$$

Then, since $$\omega_{0i}^2/\Omega_i^2 = n^2 \gg \omega_{0e}^2/\Omega_e^2,$$

$$W_{Ki} = W_H \gg W_{Ke}$$

The energy in the wave is thus shared between the ions and the magnetic field: although it might therefore be called an ionomagnetic wave, it is usually called hydromagnetic for the following reason.

The refractive index of the wave is given by

$$n = \omega_{0i}/\Omega_i = (Nm_i/\epsilon_0 B^2)^{\frac{1}{2}} \tag{8.41}$$

and, if the mass-density of the ions is written $\rho = Nm_i$, the velocity (V_A) of the wave is seen, from (8.41) to be

$$V_A = c/n = (B^2/\mu_0\rho)^{\frac{1}{2}} \tag{8.42}$$

This velocity is the same as that deduced by Alfvén [2] for the wave that travels in a conducting liquid of density ρ in the presence of a magnetic field, except that in the ionosphere $\rho = Nm_i$ represents the density of the ions alone, and does not include that of the surrounding neutral gas. Alfvén called his wave 'hydromagnetic' and it is usual to give that name to ionomagnetic waves of the type discussed here.

8.5.4 Cross-over frequencies [95]

The curves of fig. 8.3.(*c*), derived from (8.34) represent a situation in which the electrons are accompanied by an equal number of positive ions all with the same mass. In the ionosphere there are sometimes ions of more than one kind, and then it can happen that at certain frequencies, known as cross-over frequencies, the two characteristic waves have the same velocity and polarization, so that energy can be passed from one to the other. A situation of that kind is involved in the production of the naturally occurring phenomenon known as electron–ion whistlers: analyses of these whistlers has led to knowledge of the ion content of the ionosphere (see § 9.7.3).

To understand the situation suppose that there are positive ions denoted by (1) and (2) with masses m_1 and m_2 and concentrations N_1 and N_2 such that $N_1 + N_2 = N_e$, and write

$$X_1 = \frac{N_1 e^2}{\epsilon_0 m_1 \omega^2} = \frac{N_1}{N_e} \frac{m_e}{m_1} X_e$$

$$\equiv \frac{A_1}{M_1} X_e$$

where

$$A_1 = N_1/N_e \tag{8.43}$$

and

$$M_1 = m_1/m_e \tag{8.44}$$

and similarly for X_2:

we can also write

$$Y_1 = Y_e/M_1 \quad Y_2 = Y_e/M_2$$

The argument of the preceding paragraphs shows that

$$P(\text{total}) = P_e + P_1 + P_2$$

6-2

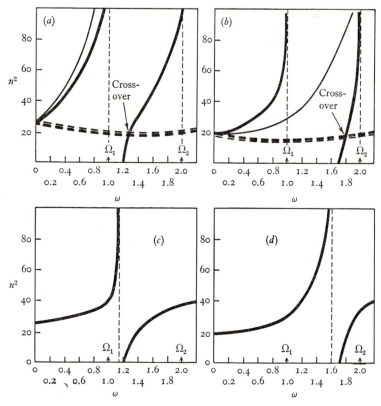

Fig. 8.5. To illustrate the occurrence of 'cross-over frequencies' when a wave travels along the direction of a magnetic field through a plasma that contains two kinds of ion with different masses. It is supposed that there are electrons with $\omega_{0e} = 10$, $\Omega_e = 4$ and ions (1) with $\Omega_1 = 1$ and (2) with $\Omega_2 = 2$: the fractional abundances of the ions are A_1 and A_2.

(a) and (b) show the dispersion curves for longitudinal propagation. In (a) the thick lines correspond to a situation where $A_1 = 0.8$ and $A_2 = 0.2$. The thin lines correspond to $A_1 = 1$ and $A_2 = 0$ for comparison. In (b) the thick lines correspond to a situation where $A_1 = 0.2$, $A_2 = 0.8$. The thin lines correspond to $A_1 = 0$ and $A_2 = 1$ for comparison. The cross-over frequency depends on the ratio A_1/A_2.

(c) and (d) show the corresponding curves for the extraordinary wave, when propagation is transverse, derived respectively from (a) and (b) by making use of the expression (8.74).

$$n^{-2} \text{ (extraordinary, transverse)} = \tfrac{1}{2} \left\{ \begin{array}{c} n^{-2} \text{ (upper sign, longitudinal)} \\ + \\ n^{-2} \text{ (lower sign, longitudinal)} \end{array} \right\}$$

so that $(n^2 - 1)$ is derived by adding the separate contributions made by the electrons and by both kinds of ions: and hence

$$n^2 - 1 = -\left(\frac{X_e}{1 \pm Y_e} + \frac{X_1}{1 \mp Y_1} + \frac{X_2}{1 \mp Y_2}\right) \tag{8.45}$$

When n^2 is plotted against ω the curves then take the forms shown in fig. 8.5: in the continuous curve, corresponding to the upper sign, there are now two frequencies Ω_1 and Ω_2 that make $n^2 = \infty$. Moreover the dashed curve for the lower sign now crosses that for the upper sign at the *cross-over frequency* ω_{co}.

If, at the frequency ω_{co} not only the refractive indices, but also the polarizations, of the two characteristic waves were the same, then one wave would excite the other as it travelled. In fact, however, for waves with their wave normals precisely in the direction of the magnetic field, like those considered here, the polarizations are not the same at the cross-over frequency. For wave-normal directions quite close to the field direction the polarizations, and the refractive indices of the two characteristic waves, are, however, very nearly the same at places where $\omega = \omega_{co}$, and one wave excites the other as it travels.

At the cross-over frequency (8.45) shows that

$$\frac{X_e}{1 + Y_e} + \frac{X_1}{1 - Y_1} + \frac{X_2}{1 - Y_2} = \frac{X_e}{1 - Y_e} + \frac{X_1}{1 + Y_1} + \frac{X_2}{1 + Y_2}$$

or

$$\frac{X_e Y_e}{1 - Y_e^2} = \frac{X_1 Y_1}{1 - Y_1^2} + \frac{X_2 Y_2}{1 - Y_2^2} \tag{8.46}$$

In the ionosphere $Y_e^2 \gg 1$ at the cross-over so that (8.46) can be written

$$\frac{1}{Y_e^2} = \frac{X_1 Y_1}{X_e Y_e}\left(\frac{1}{Y_1^2 - 1}\right) + \frac{X_2 Y_2}{X_e Y_e}\left(\frac{1}{Y_2^2 - 1}\right)$$

If use is made of (8.43) and (8.44) this can be written

$$\frac{1}{Y_{co}^2} = \frac{A_1}{Y_{co}^2 - M_1^2} + \frac{A_2}{Y_{co}^2 - M_2^2} \tag{8.47}$$

where

$$Y_{co} = \Omega_e / \omega_{co} \tag{8.48}$$

The cross-over frequency (ω_{co}) is thus independent of the electron concentration: it depends only on the relative proportions (A_1, A_2) and masses (M_1, M_2) of the ions and on the electron gyro-frequency (Ω_e). Fig. 8.5(a) corresponds to a situation where the two ions are present with relative proportions $A_1 = 0.8, A_2 = 0.2$; fig. 8.5(b) corresponds to $A_1 = 0.2, A_2 = 0.8$.

If there are ions of more than two types a similar discussion shows that the dispersion curves cross at more than one frequency, whose magnitudes can be deduced in the way outlined above.

8.6 Transverse propagation in a magnetic field

8.6.1 Only electrons moving

Consider next the situation where the wave normal is perpendicular to the superimposed magnetic field $(B_L = 0, B_T = B)$ and where electrons alone can move. The wave then travels across the field and the situation is called one of *transverse propagation*. The constitutive relations for electrons become

$$\epsilon_0 X_e E_x = -P_x - i Y_e P_z \tag{8.49}$$

$$\epsilon_0 X_e E_y = -P_y \tag{8.50}$$

$$\epsilon_0 X_e E_z = -P_z + i Y_e P_x \tag{8.51}$$

One characteristic wave solution is $E_x = E_z = P_x = P_z = 0$ and $P_y/\epsilon_0 E_y = -X_e$, so that $n^2 = 1 - X_e$. For this wave the electrons move only along the direction of the superimposed magnetic field, and the refractive index takes the form it would have in the absence of a field. It is called the *ordinary wave*, its refractive index varies as shown by the continuous lines in figs. 8.6 and 8.7.

The second, more interesting, characteristic wave has $E_y = P_y = 0$; it is called the *extraordinary wave*. The other variables satisfy the two constitutive relations (8.49) and (8.51) and the quite general relation

$$\epsilon_0 E_z = -P_z \tag{8.52}$$

Combination of (8.51) and (8.52) shows that

$$P_z/P_x = +i Y_e/(1 - X_e) \tag{8.53}$$

and insertion of this into (8.49) yields

$$\frac{P_x}{\epsilon_0 E_x} = -\frac{X_e}{1 - Y_e^2(1 - X_e)^{-1}} \tag{8.54}$$

and hence
$$n^2 = 1 - \frac{X_e}{1 - Y_e^2(1 - X_e)^{-1}} \qquad (8.55)$$

So that this expression can be compared with (8.77) appropriate to a situation where both ions and electrons can move, it is sometimes convenient to write it

$$n^2 = \frac{\{X_e - (1 + Y_e)\}\{X_e - (1 - Y_e)\}}{(1 - Y_e^2) - X_e} \qquad (8.56)$$

If n^2 is plotted against $X_e (= \omega_{0e}^2/\omega^2)$ with ω (and hence Y_e) held constant, the curves take different forms according as $Y_e >$ or < 1. Examples for $Y_e = \frac{1}{2}$ and $Y_e = 2$ are shown by the dashed lines in fig. 8.6.

To plot n^2 against ω with ω_{0e}^2 held constant we note from (8.56) that

$$\begin{aligned} n^2 = 1 \quad &\text{when } X_e = 1 \quad \text{i.e. when } \omega = \omega_{0e} \\ &\text{when } X_e = 0 \quad \text{i.e. when } \omega \to \infty \end{aligned} \right\} \qquad (8.57)$$

or

$$n^2 = \infty \quad \text{when} \quad X_e + Y_e^2 = 1 \quad \text{i.e. when}$$

$$\omega^2 = \omega_{3e}^2 = \omega_{0e}^2 + \Omega_e^2 \qquad (8.58)$$

$$n^2 = 0 \quad \text{when} \quad X_e = 1 \pm Y_e \quad \text{i.e. when}$$

$$\begin{aligned} \omega = \omega_{1e} &= (\omega_{0e}^2 + \tfrac{1}{4}\Omega_e^2)^{\frac{1}{2}} - \tfrac{1}{2}\Omega_e \\ \omega = \omega_{2e} &= (\omega_{0e}^2 + \tfrac{1}{4}\Omega_e^2)^{\frac{1}{2}} + \tfrac{1}{2}\Omega_e \end{aligned} \right\} \qquad (8.59)$$

or

The frequencies of (8.59) are the same as those of (8.29) appropriate to longitudinal propagation. The dashed line in fig. 8.7(a) shows the form of the curve $n^2(\omega)$ for the extraordinary wave when $\omega_{0e} = 10$ and $\Omega_e = 4$. There are two pass-bands, one containing all frequencies greater than ω_{2e} and one containing frequencies between ω_{3e} and ω_{1e}. The refractive index is less than unity throughout the first pass-band; in the second pass-band it is less than unity when $\omega < \omega_{0e}$ and greater when $\omega > \omega_{0e}$: at the high frequency limit of the band it is infinite.

It is interesting to consider the movements of the electrons (and the associated volume polarization P) in the extraordinary wave. In it $P_y = 0$ and from (8.53) $P_z/P_x = i Y_e/(1 - X_e)$ so that, since the polarization is proportional to the displacement of an electron, the electrons move in ellipses with axes along the x- and z-directions and with ratio

$(1 - X_e)/Y_e$: the direction of rotation is different according as $X_e >$ or < 1. When $X_e = 1$ the electron moves in a straight line along the z-direction, $P_x = 0$ and hence $n^2 = 1$. Equation (8.54) shows that, as X_e passes through the value unity P_x/E_x changes sign: when $X_e > 1$ $(\omega^2 < \omega_{0e}^2)$ it is negative and when $X_e < 1$ $(\omega^2 > \omega_{0e}^2)$ it is positive.

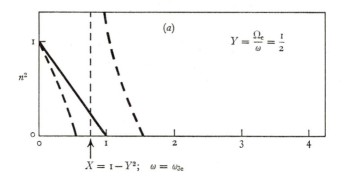

$$(a) \qquad Y = \frac{\Omega_e}{\omega} = \frac{1}{2}$$

$$X = 1 - Y^2; \quad \omega = \omega_{3e}$$

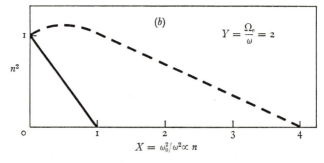

$$(b) \qquad Y = \frac{\Omega_e}{\omega} = 2$$

$$X = \omega_0^2/\omega^2 \propto n$$

Fig. 8.6. $n^2(X)$ curves for transverse propagation in a situation where only electrons move. The continuous line represents the ordinary wave $(n^2 = 1 - X)$. The dashed line represents the extraordinary (8.55 or 8.56), it takes different forms according as $Y <$ or > 1.

These movements of the electron correspond to the fact, noticeable in the pass-band where $\omega_{1e} < \omega < \omega_{3e}$ in fig. 8.7(a) that $n^2 > 1$ when $\omega > \omega_{0e}$ and $n^2 < 1$ when $\omega < \omega_{0e}$.

The oscillations of the electrons in the z-direction produce alternate compressions and rarefactions of the electron concentration, similar to those in a sound wave, and the resulting space charge produces the longitudinal component E_z of the field. When we discuss the transverse propagation of waves through a mixture of ions and electrons we shall see that their movements are coupled by the field of a similar space charge.

When $\omega = \omega_{3e}$ the refractive index becomes infinite. P_x/E_x is then infinite, so that E_x is zero and the only finite field is E_z: this field, by itself, derives the electron in its elliptical orbit. The situation can be discussed more fully in the following way. A field E_z will always drive an electron in an elliptical orbit in the zx plane whatever the frequency: the situation is described by setting $E_x = E_y = 0$ in (8.49), (8.50) and (8.51) which then become

$$0 = -P_x - i\, Y_e P_z \qquad (8.60)$$

$$\epsilon_0 X_e E_z = -P_z + i\, Y_e P_x \qquad (8.61)$$

where (8.60) shows that the polarization P, and hence the displacement of the electron, is represented by an ellipse in the zx plane given by

$$P_x/P_z = -i\, Y_e \qquad (8.62)$$

Insertion of (8.62) into (8.61) shows that

$$P_z = -\frac{X_e}{1 - Y_e^2} \epsilon_0 E_z \qquad (8.63)$$

This relation applies at all frequencies. If however a characteristic wave is to travel along the z-direction then, as usual, the relation $P_z = -\epsilon_0 E_z$ must also be satisfied: this can happen only when

$$\frac{X_e}{1 - Y_e^2} = 1 \qquad (8.64)$$

We thus arrive at the relation (8.58) that applies when the refractive index is infinite.

Since this result is also applicable to the more general situation where both ions and electrons can move (see p. 164), it is useful to restate it as follows:

An oscillating electric field E_z entirely along the wave-normal direction, will in general produce an elliptical variation of P in the zx plane, with axes P_x and P_z. If in addition $\epsilon_0 E_z = -P_z$ then a characteristic wave can travel and since there is no field E_x, $P_x/E_x = \infty$, and the refractive index is infinitely great.

8.6.2   Ions and electrons both moving

Suppose next that both ions and electrons move under the influence of the wave, and that the wave normal is again perpendicular to the imposed magnetic field (transverse propagation). The equations of motion (8.12, 13, 14, (c) and (d)) for the electrons and for the ions then become:

Electrons

$$\epsilon_0 X_e E_x = -P_{xe} - i Y_e P_{ze} \tag{8.65}$$

$$\epsilon_0 X_e E_y = -P_{ye} \tag{8.66}$$

$$\epsilon_0 X_e E_z = -P_{ze} + i Y_e P_{xe} \tag{8.67}$$

Ions

$$\epsilon_0 X_i E_x = -P_{x1} + i Y_i P_{z1} \tag{8.68}$$

$$\epsilon_0 X_i E_y = -P_{y1} \tag{8.69}$$

$$\epsilon_0 X_i E_z = -P_{z1} - i Y_i P_{x1} \tag{8.70}$$

There is also the general condition, for a characteristic wave that

$$\epsilon_0 E_z = -P_z(\text{total}) = -P_{ze} - P_{z1} \tag{8.71}$$

This condition provides coupling between the motions of the ions and of the electrons: the situation for transverse propagation is thus more complicated than that for longitudinal propagation where there is no such coupling.

One characteristic wave solution is

$$E_x = E_z = P_{xe} = P_{x1} = P_{ze} = P_{z1} = 0$$

together with

$$P_{ye} = -\epsilon_0 X_e E_y \quad \text{and} \quad P_{y1} = -\epsilon_0 X_i E_y$$

Then

$$P_y(\text{total})/\epsilon_0 E_y = -(X_e + X_i)$$

so that

$$n^2 = 1 - (X_e + X_i) = 1 - X \tag{8.72}$$

The dispersion relation (8.72) describes a characteristic wave in which the electric field (E_y) is along the direction of the imposed magnetic field, the wave therefore behaves as though there were no imposed field: it is called the ordinary wave. The second characteristic wave is called the extra-ordinary wave, its dispersion relation is derived from (8.65)–(8.71) by putting $E_y = P_{ye} = P_{y1} = 0$; and after a little algebraic manipulation there results

$$n^2 = \frac{[X-(1+Y_e)(1-Y_i)][X-(1-Y_e)(1+Y_i)]}{(1-Y_e^2)(1-Y_i^2)+X(Y_e Y_i-1)} \quad (8.73)$$

If (8.73), that gives the square of the refractive index (n_E^2), for the extraordinary wave when propagation is perpendicular to the magnetic field is compared with (8.34) that gives the squares of the refractive indices (n_u^2) and (n_i^2) for the waves corresponding to the upper and lower signs when the propagation is along the field, it is seen that

$$\frac{1}{n_E^2} = \frac{1}{2}\left(\frac{1}{n_u^2}+\frac{1}{n_i^2}\right) \quad (8.74)$$

Stix [30] has demonstrated the truth of (8.74) from quite general first principles. It is a particularly valuable result because it allows the dispersion relations for the extraordinary transverse wave to be deduced at once from those for the two longitudinal waves. Thus, for example, the curves of fig. 8.7 can be deduced from those of fig. 8.3 (extended to negative values of n^2).

To plot the dispersion curve for the extraordinary transverse wave consider first the value of n^2 when the frequency is very small. It is sufficient to make use of (8.74) and to recall that, for longitudinal propagation, the two refractive indices are equal, and are given by (8.38); (8.74) then shows that the refractive index for the extraordinary transverse wave must have the same value, i.e.

$$n^2 = 1 + \omega_0^2/\Omega_e\Omega_i \quad (8.75)$$

This same result can also be obtained from (8.73) for a situation where $Y_e \ll 1$ and $Y_i \ll 1$.

To help in plotting the dispersion curves for the extraordinary wave (8.73) can be used to show that the refractive index has the important values 0, 1 and ∞ when the following relations are satisfied.

$$n^2 = 0 \quad \text{when} \quad X = (1+Y_e)(1-Y_i) \quad \text{or} \quad (1-Y_e)(1+Y_i) \quad (8.76)$$

$$n^2 = 1 \quad \text{when} \quad X = 1 - Y_e Y_i \quad \text{or} \quad X = 0 \quad (8.77)$$

$$n^2 = \infty \quad \text{when} \quad X = \frac{(1-Y_e^2)(1-Y_i^2)}{1-Y_e Y_i} \quad (8.78)$$

These expressions show that the frequencies that make $n^2 = 0$, 1, and ∞ are as follows:

$$n^2 = 1 \quad \text{when} \quad \omega = [\omega_0^2 + \Omega_e \Omega_i]^{\frac{1}{2}} \tag{8.79}$$

$$\text{or} \quad \omega = \infty$$

$$n^2 = 0 \quad \text{when} \quad \omega = \omega_1 = [\omega_0^2 + \tfrac{1}{4}(\Omega_e + \Omega_i)^2]^{\frac{1}{2}} - \tfrac{1}{2}[\Omega_e - \Omega_i] \tag{8.80}$$

$$\text{or} \quad \omega = \omega_2 = [\omega_0^2 + \tfrac{1}{4}(\Omega_e + \Omega_i)^2]^{\frac{1}{2}} + \tfrac{1}{2}[\Omega_e - \Omega_i] \tag{8.81}$$

$$n^2 = \infty \quad \text{when}$$

$$\omega = \omega_3 = \left[\tfrac{1}{2}(\omega_0^2 + \Omega_e^2 + \Omega_i^2) + \tfrac{1}{2}\{(\omega_0^2 + \Omega_e^2 + \Omega_i^2)^2 - 4\Omega_e \Omega_i(\omega_0^2 + \Omega_e \Omega_i)\}^{\frac{1}{2}} \right]^{\frac{1}{2}} \tag{8.82}$$

$$\text{or}$$

$$\omega = \omega_4 = \left[\tfrac{1}{2}(\omega_0^2 + \Omega_e^2 + \Omega_i^2) - \tfrac{1}{2}\{(\omega_0^2 + \Omega_e^2 + \Omega_i^2)^2 - 4\Omega_e \Omega_i(\omega_0^2 + \Omega_e \Omega_i)\}^{\frac{1}{2}} \right]^{\frac{1}{2}} \tag{8.83}$$

If $\Omega_i = 0$ these expressions for ω_1, ω_2 and ω_3 are the same as (8.58) and (8.59) for ω_{1e}, ω_{2e} and ω_{3e} appropriate to a situation where only electrons move. The frequency ω_4 becomes zero when $\Omega_i = 0$: there is no corresponding frequency when only electrons can move.

Fig. 8.7 shows dispersion curves represented by (8.73) for a medium containing various combinations of electrons with $\omega_{0e} = 10$, $\Omega_e = 4$, and ions with $\omega_{0i} = 5$, $\Omega_i = 1$, and $m_i/m_e = 4$. The three sets of curves correspond to situations where

(a) only electrons can move ($\omega_{0e} = 10$, $\Omega_e = 4$ and $\omega_{0i} = \Omega_i = 0$),
(b) only ions can move ($\omega_{0i} = 5$, $\Omega_i = 1$ and $\omega_{0e} = \Omega_e = 0$),
(c) both electrons and ions can move ($\omega_{0e} = 10$, $\omega_{0i} = 5$, $\Omega_e = 4$, $\Omega_i = 1$).

In the ionosphere $m_i/m_e = 10^4$ so that ω_0, ω_1, ω_2, ω_3 are nearly the same as ω_{0e}, ω_{1e}, ω_{2e}, ω_{3e} and for frequencies greater than ω_{1e} the curves (a) for electrons alone moving are nearly the same as those (c) for electrons and ions both moving. When the ions can move there is, however, a new finite frequency ω_4 at which the refractive index becomes infinite. At frequencies less than this there is a pass-band, in which n^2 is positive and waves can travel: there is no such pass-band if the ions cannot move. The frequency ω_4, that limits this pass-band at its upper end, is determined jointly by the movements of the

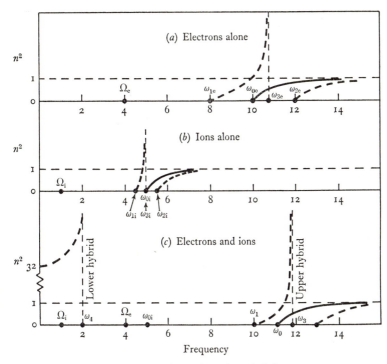

Fig. 8.7. Dispersion curves (n^2 as a function of ω) for transverse propagation drawn for $m_i/m_e = 4$, $\omega_{0e} = 10$, $\omega_{0i} = 5$, $\Omega_e = 4$, $\Omega_i = 1$: these values are very different from the ionospheric ones. The continuous lines correspond to (8.72) and the dashed lines to (8.73) and to situations where (a) only electrons can move ($\omega_{0i} = \Omega_i = 0$); (b) only ions can move ($\omega_{0e} = \Omega_e = 0$); (c) electrons and ions can both move. Unlike the corresponding curves for longitudinal propagation (fig. 8.3) curve (c) is *not* the sum of curves (a) and (b).

electrons and the ions, it is called the *lower hybrid frequency*. The other frequency ω_3 that also makes $n^2 = \infty$ is called the *upper hybrid frequency*. Although it, also, is determined jointly by the movements of the ions and electrons, the effect of the ions is slight.

8.6.3 The upper and lower hybrid frequencies

The properties of the ionosphere at the two hybrid frequencies have been used in ionospheric investigations (chapters 9 and 10) and it is of interest to examine them in more detail.

A discussion of the motions of the ions and electrons that make $n^2 = \infty$ at the upper and lower hybrid frequencies is similar to that of §8.6.1 for the situation where only electrons can move. For the wave under discussion $E_y = P_{ye} = P_{yi} = 0$ and when

$$n^2 \to \infty, \quad P_x(\text{total})/E_x \to \infty$$

so that $E_x \to 0$ and the electric field is entirely along the z-direction. Then, as in (8.63)

$$P_{ze} = -\frac{X_e}{1 - Y_e^2}\epsilon_0 E_z \tag{8.84}$$

$$P_{zi} = -\frac{X_i}{1 - Y_i^2}\epsilon_0 E_z \tag{8.85}$$

so that $\quad P_z(\text{total}) = P_{ze} + P_{zi} = -\left(\frac{X_e}{1 - Y_e^2} + \frac{X_i}{1 - Y_i^2}\right)\epsilon_0 E_z \tag{8.86}$

But if a characteristic wave is to travel with wave normal along Oz the general condition

$$P_z(\text{total}) = -\epsilon_0 E_z \tag{8.87}$$

must also apply; this can happen only if

$$\frac{X_e}{1 - Y_e^2} + \frac{X_i}{1 - Y_i^2} = 1 \tag{8.88}$$

The condition (8.88) that makes $n^2 = \infty$, corresponds to (8.64) for moving electrons and is equivalent to (8.78).

Each term on the left-hand side of (8.88) changes sign as the magnitude of the appropriate Y (Y_e or Y_i) passes through unity. If both Y_e and Y_i are greater than unity ($\omega < \Omega_i < \Omega_e$) both terms are negative and the equation cannot be satisfied for any value of ω: if, however, $Y_e > 1$ and $Y_i < 1$ (i.e. $\Omega_i < \omega < \Omega_e$) the two terms are of opposite sign and there is a value of ω that satisfies (8.88). Equations (8.84) and (8.85) then show that the z-components of polarization are in anti-phase so that the ions and electrons, with charges of opposite sign, have displacements that are in phase. *This is the situation at the lower hybrid frequency.*

If $Y_e < 1$ and $Y_i < 1$ ($\Omega_i < \Omega_e < \omega$) both terms on the left-hand side are positive and a second solution is possible in which P_{ze} and P_{zi} are in the same phase (the particles being displaced in opposite senses). This is the situation at the *upper hybrid frequency.*

In the ionosphere the lower hybrid frequency ω_4 is usually such that $\Omega_i^2 \ll \omega_4^2 \ll \Omega_e^2$; thus for example

$$\omega_4 \fallingdotseq 2\pi \times 10^4, \quad \Omega_i \fallingdotseq 2\pi \times 10^2 \quad \text{and} \quad \Omega_e \fallingdotseq 2\pi \times 10^6 \, \text{s}^{-1}.$$

Then with $X_{4e} = \omega_{0e}^2/\omega_4^2$, $X_{4i} = \omega_{0i}^2/\omega_4^2$ and $Y_{4e} = \Omega_e/\omega_4$ (8.88) can be written

$$X_{4i} = 1 + \frac{X_{4e}}{Y_{4e}^2} \tag{8.89}$$

or

$$\frac{m_e}{m_i} \frac{1}{\omega_4^2} = \frac{1}{\omega_{0e}^2} + \frac{1}{\Omega_e^2} \tag{8.89a}$$

This expression leads to the possibility of measuring the mass (m_i) of the ions in the ionosphere surrounding a satellite by measuring the lower hybrid frequency (ω_4), the plasma frequency (ω_{0e}) and the electron gyro-frequency (Ω_e). The discussion can be extended, as follows, to apply to a situation where more than one kind of positive ion is present.

Suppose several positive ions of different masses $m_1, m_2, ..., m_j$ are present with concentrations $A_1 N_e, A_2 N_e, ..., A_j N_e$ and write

$$X_{4j} = \frac{A_j N_e e^2}{\epsilon_0 \omega_4^2 m_j} = X_{4e} m_e (A_j/m_j)$$

and

$$1/m_{\text{eff}} = \Sigma_j (A_j/m_j) \tag{8.90}$$

Then (8.88) is replaced by

$$\frac{X_e}{1 - Y_e^2} + \Sigma_j \left(\frac{X_{j4}}{1 - Y_j^2} \right) = 1 \tag{8.91}$$

and, with the same approximations as before (8.89) is replaced by

$$\Sigma X_{4j} = X_{4e}(m_e/m_{\text{eff}}) = 1 + X_{4e}/Y_{4e}^2$$

and (8.89a) by

$$\frac{m_e}{m_{\text{eff}}} \frac{1}{\omega_4^2} = \frac{1}{\omega_{0e}^2} + \frac{1}{\omega_e^2} \tag{8.92}$$

The use of (8.92) to deduce the effective mass m_{eff} (defined by (8.90)) from satellite measurements of the lower hybrid frequency ω_4 is discussed in §9.7.5.

When more than one kind of ion is present the dispersion curve for frequencies less than the lower hybrid frequency (ω_4) takes a form that can be deduced from the curves for the two longitudinal waves by the use of (8.74). Examples are shown in fig. 8.5 (*c*) and (*d*). In making use of (8.74) in this way to construct the curves of (*c*) from those of (*a*) and (*d*) from those of (*b*) it is useful to notice that

$$n_E^2 = \infty \quad \text{when} \quad n_u^2 = -n_l^2$$

$$n_E^2 = 0 \quad \text{when} \quad n_u^2 = 0 \quad \text{or} \quad n_l^2 = 0$$

$$n_E^2 = 2n_l^2 \quad \text{when} \quad n_u^2 = \infty$$

$$n_E^2 = 2n_u^2 \quad \text{when} \quad n_l^2 = \infty$$

$$n_E^2 = n_u^2 = n_l^2 \quad \text{when} \quad n_u^2 = n_l^2 \text{ (at the cross-over frequency).}$$

8.7 Electro-acoustic waves [8 (p. 176), 15]

In the preceding discussions forces arising from partial pressure gradients were omitted: the only forces acting on the electrons and ions were electrical in nature. But the forces of partial pressure give rise to sound waves in a neutral gas, and it is interesting to discuss the part they play in a plasma. It will be shown that waves are possible in which the particles move only in the wave normal direction, as in sound waves: they are called *plasma waves* or *electro-acoustic* waves, or sometimes *electron-acoustic, ion-acoustic, or electron–ion-acoustic waves* to indicate their nature more precisely. They can travel only in a hot plasma where the thermal motions provide the necessary partial-pressure gradients.

A detailed theory involves discussion of the nature of partial pressure when collisions are infrequent, but for the sake of simplicity these considerations are omitted here, and it is supposed that the pressures act as in a gas that transmits sound waves. A complete treatment also recognizes the adiabatic nature of the wave compressions, but the final result is the same as that arrived at here.

In the elementary theory of a sound wave travelling in the z-direction in a neutral gas containing N_0 molecules in unit volume it is supposed that the molecules originally on a plane at z are displaced a distance ζ in the z-direction, those originally at $z + \Delta z$ being displaced

$$\zeta + (\partial \zeta / \partial z) \Delta z$$

The concentration at a distance z then becomes $N_0 + N_1$ where $N_1 = -N_0(\partial \zeta / \partial z)$ and the partial pressure, which was $p_0 (= N_0 m v^2)$

before the wave existed, becomes p_0+p_1 where $p_1 = N_1mv^2$, v being the r.m.s. gas-kinetic velocity. The equation of motion of a molecule at z is then

$$m\frac{\partial^2\zeta}{\partial t^2} = -\frac{1}{N_0}\frac{\partial p_1}{\partial z} = mv^2\frac{\partial^2\zeta}{\partial z^2} \qquad (8.93)$$

The displacement ζ of the molecules thus travels as a wave-motion along the z-direction with velocity $V = v$, the r.m.s. gas-kinetic velocity.

Now suppose that the particles are charged, so that a change in their concentration produces bunching in the z-direction accompanied by a space charge $\rho = eN_1$ and a field E_z given by $\rho = \epsilon_0(\partial E_z/\partial z)$. Hence

$$\epsilon_0\frac{\partial E_z}{\partial z} = eN_1 = -eN_0\frac{\partial\zeta}{\partial z}$$

or

$$\epsilon_0 E_z = -eN_0\zeta \qquad (8.94)$$

The equation of motion (8.93) of a charged particle then becomes

$$m\frac{\partial^2\zeta}{\partial t^2} = E_z e + mv^2\frac{\partial^2\zeta}{\partial z^2} \qquad (8.95)$$

By substituting for E_z from (8.94) and by writing $\zeta = \zeta_0\exp(i\omega t)$, (8.95) can be written

$$-\omega^2\left(1 - \frac{N_0 e^2}{\epsilon_0 m\omega^2}\right)\zeta_0 = v^2\frac{\partial^2\zeta_0}{\partial z^2}$$

hence

$$\zeta_0 = A\exp\left[-i\{(\omega/v)(1-X)^{\frac{1}{2}}z\}\right]$$

where

$$X = N_0 e^2/\epsilon_0 m\omega^2$$

and then

$$\xi = A\exp\left[i\omega(t - \{(1-X)^{\frac{1}{2}}/v\}z)\right]$$

Hence the particle displacement ζ travels as a wave motion with velocity V given by

$$V = v(1-X)^{-\frac{1}{2}} \qquad (8.96)$$

This result can also be derived by a modification of the approach used previously in discussing electromagnetic waves. Suppose that a wave can travel as represented by $\exp i(\omega t - kx)$ so that (8.95) becomes

$$E_z e = -m\omega^2\zeta + mv^2 k^2\zeta \qquad (8.97)$$

Now write

$$Ne\zeta = P_z \quad \text{and} \quad \omega/k = V$$

to give
$$E_z e = -\frac{m\omega^2}{Ne}(1 - v^2/V^2)P_z$$

or
$$P_z = -\epsilon_0 E_z \frac{X}{1 - v^2/V^2} \qquad (8.98)$$

But if a wave involving an electric field E_z is to travel along the z-direction the general relation $P_z = -\epsilon_0 E_z$ must be obeyed, hence

$$X = 1 - v^2/V^2 \qquad (8.99)$$

in agreement with the previous result.

If not only electrons (subscript e), but also ions (subscript i) take part in the wave motion, their equation of motion is similar to that of the electrons, and corresponding to (8.98)

$$\left. \begin{aligned} P_{ze} &= \frac{-X_e}{1 - v_e^2 V^2}\epsilon_0 E_z \\[2mm] P_{zi} &= \frac{-X_i}{1 - v_i^2/V^2}\epsilon_0 E_z \end{aligned} \right\} \qquad (8.100)$$

and

so that
$$P_z(\text{total}) = P_{ze} + P_{zi}$$
$$= -\left(\frac{X_e}{1 - v_e^2/V^2} + \frac{X_i}{1 - v_i^2/V^2}\right)\epsilon_0 E_z \qquad (8.101)$$

But, since the general relation $P_z(\text{total}) = -\epsilon_0 E_z$ must hold, it follows that

$$\frac{X_e}{1 - v_e^2/V^2} + \frac{X_i}{1 - v_i^2/V^2} = 1 \qquad (8.102)$$

and this is the dispersion relation that gives the wave velocity V in terms of the frequency ω (contained in $X_e = \omega_{0e}^2/\omega^2$ and $X_i = \omega_{0i}^2/\omega^2$).

Without solving (8.102) in full it is easy to see some important points that are useful in discussing the incoherent scatter of waves in the ionosphere (p. 192). First we see that a solution in which $V^2 \gg v_e^2$ (and hence $\gg v_i^2$) requires that

$$X_e(1 + v_e^2/V^2) + X_i(1 + v_i^2/V^2) \fallingdotseq 1$$

or, since $\quad v_e^2 X_e = (m_i/m_e)^2 v_i^2 X_i \quad$ and $\quad m_i/m_e \gg 1$,

$$V^2 \fallingdotseq v_e^2 \frac{X_e}{1 - X_e - X_i}$$

$$\fallingdotseq v_e^2 \frac{X}{1 - X} \qquad (8.103)$$

This wave has infinite velocity when $X = 1$ ($\omega = \omega_0$); it then corresponds to the bodily oscillation of the plasma at the plasma frequency, deduced in § 8.4 where partial pressures were neglected.

Equation (8.101) shows that when

$$V^2 \gg v_e^2 \text{ and } \gg v_i^2, \quad P_{ze}/P_{zi} = X_e/X_i \gg 1,$$

hence the displacement of the electrons is much greater than that of the ions: the wave is therefore called an electron-acoustic wave. Since P_{ze} and P_{zi} have the same sign, whereas the charges have opposite sign, the displacements of the electrons and the ions are in opposite directions.

Next we seek a solution for which $V^2 \ll v_e^2$, so that (8.102) becomes

$$\frac{X_i}{1 - v_i^2/V^2} - \frac{X_e}{v_e^2/V^2} = 1 \tag{8.104}$$

It is convenient to discuss a situation where the electron and ion temperatures, T_e and T_i are different, so that $v_e^2/v_i^2 = \alpha(m_i/m_e)$ where $\alpha = T_e/T_i$. Then

$$X_i/v_i^2 = \alpha X_e/v_e^2 \tag{8.105}$$

and (8.104) becomes

$$\frac{1}{1 - v_i^2/V^2} - \frac{1}{\alpha v_i^2/V^2} = \frac{1}{X_i} \tag{8.106}$$

If $X_i = 0$ (8.106) is satisfied when $V^2 = v_i^2$, and if $X_i = \infty$ it is satisfied by $V^2 = (1 + \alpha) v_i^2$; for other values of X_i the value of V lies between these two. The velocity of the wave is thus of the same order as the r.m.s. velocity of the ions: a wave of this kind plays an important part in the incoherent scatter of radio waves in the ionosphere.

The two terms on the left of (8.106) are proportional to P_{zi} and P_{ze} so that

$$\frac{P_{zi}}{P_{ze}} = -\frac{\alpha v_i^2/V^2}{1 - v_i^2/V^2} \tag{8.107}$$

and, with $(1 + \alpha) > V^2/v_i^2 > 1$, this ratio is always negative so that P_{ze} and P_{zi} are in anti-phase and the displacements of the electrons and ions are in phase.

8.7.1 Interaction of waves and particles. Landau and cyclotron damping

Most of the electrons and ions in the ionosphere are moving much more slowly than the electromagnetic waves; the electric field of the wave causes them to oscillate at the wave frequency, and the wavelets that they radiate combine with the original wave to alter the refractive index in the ways described earlier in this chapter. If, however, a few charged particles are moving along the wave-normal direction with a velocity that is comparable with the wave velocity they will be acted upon by a field that oscillates at a Doppler-shifted frequency which depends on the ratio between the velocities of the particles and of the wave. When this shifted frequency takes certain values the oscillating electric field can produce a cumulative effect so that the particles are either speeded up or slowed down. If they are speeded up energy is abstracted from the wave which is correspondingly attenuated; if they are slowed down energy is transferred to the wave which is correspondingly enhanced.

Two particular situations are important: (*a*) when charged particles move along the wave-normal direction of an electro-acoustic wave with nearly the wave velocity; (*b*) when charged particles move in a helix along a magnetic field at such a speed that their frequency of gyration is equal to the Doppler-shifted frequency of a circularly polarized electro-magnetic wave travelling in the same direction.

For simplicity consider an electro-acoustic wave in the absence of a magnetic field, and recall that its electric field is along the wave-normal direction. If a charged particle were moving along that direction with the same speed as the wave it would experience a constant force in the direction of its motion; the force would either accelerate it or decelerate it, according to its position in the wave, so that it would not continue to travel with the wave. Calculation of the resultant energy transfer requires careful analysis that becomes very complicated. The result has been summarized (p. 132 in reference 30) by saying that if there is a group of charged particles drifting along the wave-normal direction

the absorption of energy from the wave is large when there are many particles streaming infinitesimally slower than the field-phase-velocity and somewhat fewer particles streaming infinitesimally faster than the field-phase-velocity. If there are more fast particles than slow ones in the neighbourhood of the phase velocity, the particles will give up energy to the wave rather than absorb it.

When energy is abstracted from the wave, the resultant damping is called *Landau damping* after the theorist who first realized its importance: it is convenient to give the name Landau growth to the corresponding enhancement of the wave that can sometimes occur.

If the charged particles have a Maxwell distribution the damping is greatest when the wave velocity is equal to their r.m.s. velocity. It is thus very great for the ion-dominated electro-acoustic wave that has velocity of the same order as the r.m.s. velocity of the ions. The velocity of the electron-acoustic wave is normally much greater than the velocity of any appreciable number of electrons, and it is not damped. Sometimes, however, energetic electrons are present, travelling at nearly the same speed as the wave, and with a velocity distribution such that the number travelling more rapidly is greater than the number travelling more slowly; there is then Landau growth and the wave amplitude increases. Damping of the ion-dominated wave and enchancement of the electron-dominated wave are important in incoherent scattering of radio waves (§9.4).

In the situation described above under (*b*) a particle moving in a helix along a constant magnetic field experiences the electric field of a circularly polarized wave rotating in synchronism with its own circular motion. The field is thus always in the same direction in relation to its own displacement so that the particle is either accelerated or decelerated and energy is abstracted from or fed to the wave. As with Landau damping a detailed calculation requires that a group of particles, with slightly different energies, should be considered. It is then found that under some circumstances the particles can give energy to the waves; some of the low-frequency waves observed in the ionosphere are supposed to have originated in this way. Under other circumstances the particles lose energy to the waves; it has been suggested that this loss provides one mechanism for the removal of protons trapped on geomagnetic field lines. The phenomena described here have been called *cyclotron damping* and *cyclotron enhancement*.

8.8 Groups, angular spectra and wave packets [3, 9]

The preceding discussions of waves were all concerned with infinite plane waves of infinite duration. It is now necessary to extend them to apply to wave packets limited in extent both across the wave front and in the direction of travel. The phenomena to be discussed are important in the formation of 'whistlers' (§ 9.7.2).

Limitation of a wave along the direction of its travel produces a 'group' with an infinitely wide wave front. As an introduction to the problem of wave packets the standard theory of groups will therefore first be developed in outline.

Suppose that a wave $f(t, z)$ of frequency ω_0, with uniform amplitude over its wave front, has its wave normal in the z-direction and that its amplitude is modulated by a short pulse $a(t)$ so that, in the plane $z = 0$ the disturbance is represented by†

$$f(t, 0) = a(t) \exp(i\omega_0 t)$$

We shall call it a *wave group*. The disturbance in the plane $z = 0$ can be Fourier-synthesized by writing

$$f(t, 0) = \int_0^\infty F(\omega) \exp(i\omega t)\, d\omega \tag{8.108}$$

where $F(\omega)$ is called the *frequency spectrum*. We shall suppose that the duration of the pulse $a(t)$ is long compared with $1/\omega_0$, so that $F(\omega)$ has a narrow maximum near $\omega = \omega_0$.

Each component in the spectrum travels as a wave motion represented by $F(\omega) \exp[i\{\omega t - k(\omega)z\}]$ and if the medium is dispersive the corresponding velocities $\omega/k(\omega)$ depend on ω. The total wave disturbance can then be written

$$f(t, z) = \int_0^\infty F(\omega) \exp[i\{\omega t - k(\omega)\}z]\, d\omega \tag{8.109}$$

Now suppose that the short pulse, $a(t)$, has its maximum at $t = 0$, so that the group can be said to pass the plane $z = 0$ at the time $t = 0$. To find the velocity of the group we find the time (t) for it to travel a distance z. When the phases of the component waves in (8.109) are stationary with respect to a change of frequency, the wave $f(t, z)$ has its maximum amplitude and the group will be found at the

† When a complex expression is written it is understood that the real part is to be taken.

corresponding time and place. Since the dominant frequency is ω_0 the phase is stationary when

$$\left[\frac{\partial}{\partial\omega}\{\omega t - k(\omega)\, z\}\right]_{\omega_0} = 0$$

or
$$t - [\partial k(\omega)/\partial\omega]_{\omega_0} z = 0 \tag{8.110}$$

The group thus travels along the wave-normal direction with a velocity

$$V_n = z/t = [\partial k(\omega)/\partial\omega]_{\omega_0}^{-1} \tag{8.111}$$

This is frequently called the group velocity, but in a discussion of an anisotropic medium like the ionosphere it is best to call it the *wave-normal group velocity* for reasons that are explained below.

Consider next a wave of a single frequency ω_0 travelling in the z-direction with amplitude constant over its wave front in the y-direction but varying over the x-direction. If the x-variation were cosinusoidal the disturbance could be synthesized by adding two equal infinite plane waves whose wave-normals make angles $\pm\theta$ with the z-direction as follows

$$\cos\{\omega_0 t - k(x\sin\theta + z\cos\theta) + \psi\}$$
$$+ \cos\{\omega_0 t - k(-x\sin\theta + z\cos\theta) - \psi\}$$
$$= 2\cos\{(k\sin\theta)\, x - \psi\}\cos\{\omega_0 t - (k\cos\theta)\, z\} \tag{8.112}$$

to produce a wave travelling along the z-direction and having a cosinusoidal amplitude variation over its wave front given by

$$\cos\{(k\sin\theta)\, x - \psi\}$$

Suppose now that in the plane $z = 0$ the amplitude varies over the x-direction as represented by a function $a(x)$ that is not cosinusoidal; the wave can be synthesized by adding several waves like those of (8.112), with cosine variations in the x-direction and with different amplitudes and phases. Each of these, in turn, can be constructed from two infinite plane waves, with suitable phases, travelling in suitable directions $\pm\theta$ as represented by

$$f(x, z, t) = \int_{-\pi}^{+\pi} F(\theta)\exp\left[i\{-k(\theta)\,(x\sin\theta + z\cos\theta)\}\right]\,d\theta \cdot \exp\left(i\omega_0 t\right) \tag{8.113}$$

This assembly of plane waves is called the *angular spectrum* that makes up the disturbance.

Suppose further that $a(x)$ is a symmetrical function with a maximum at $x = 0$, corresponding to a situation where the wave is limited by a narrow slit in the plane $z = 0$, and that, although $a(x)$ is narrow† in comparison with distances along the z-direction still to be considered, it is wide compared with the wavelength. Then, in $(8.113)\sin\theta \doteqdot \theta \ll 1$ and the angular spectrum extends over a narrow range of angles $\pm\Delta\theta$ where $\Delta\theta \ll 1$.

In discussing this angular spectrum it must be remembered that the ionosphere is an anisotropic medium and the wave velocity, and hence the wave number k, depends on the direction of propagation. There is axial symmetry round the direction of the earth's magnetic field so that the wave properties can be represented in a polar plot by a plane curve that shows how the refractive index varies with the wave-normal direction (for an example see fig. 8.8). The three-dimensional situation is then represented by the surface formed when this curve is rotated about the field direction. The shape of the curve depends on the wave frequency and the plasma frequency. The discussion of the refractive index in earlier parts of this chapter referred only to propagation along and perpendicular to the magnetic field.

The amplitude of the wave represented by (8.113) is maximum at places where the phase in the angular spectrum is stationary for changes of the angle θ around its dominant value zero, i.e. where

$$\left[\frac{\partial}{\partial\theta}\{k(\theta)(x\sin\theta + z\cos\theta)\}\right]_{\theta=0} = 0$$

or $[(\partial k/\partial\theta)(x\sin\theta + z\cos\theta) + k(x\cos\theta - z\sin\theta)]_{\theta=0} = 0$ (8.114)

The line joining this new point (x, z) to the point $(0, 0)$ where the amplitude was originally maximum thus makes an angle α with the wave-normal direction Oz given by

$$\tan\alpha = \frac{x}{z} = -\frac{1}{k}\frac{\partial k}{\partial\theta} \tag{8.115}$$

† This condition limits the discussion to distances greater than the 'Rayleigh distance' where the diffraction pattern of the slit is of the Fraunhoffer rather than the Fresnel type (see reference 126).

This direction, along which the place of maximum amplitude travels, is called the *ray direction*.

It is often useful to consider the travel of a wave disturbance that is limited both across its wave front and in its direction of travel, it is called a *wave packet*. If V_r is its velocity along the ray direction, and V_n its velocity along the wave-normal direction, then

$$V_n = V_r \cos \alpha \qquad (8.116)$$

Theorists usually call the velocity V_r, along the ray direction, the *group velocity*. This nomenclature is, however, not always used by ionospheric workers; they frequently call V_n the group velocity and V_r the ray–group velocity or the *wave packet velocity*.

It is convenient to express the ray direction and the velocity V_n in terms of the refractive index μ by using the relation $k = 2\pi/\lambda = \omega\mu/c$. Then from (8.115)

$$\tan \alpha = -\frac{1}{\mu}\frac{\partial \mu}{\partial \theta} \qquad (8.117)$$

and from (8.111)

$$\frac{1}{V_n} = \frac{\mu + \omega\,\partial\mu/\partial\omega}{c} \qquad (8.118)$$

If a *group refractive index* μ' is defined by writing $V_n = c/\mu'$ then

$$\mu' = \mu + \omega\,\partial\mu/\partial\omega \qquad (8.119)$$

Although there is some doubt about the name for the velocity V_n there seems to be no doubt that μ' should be called the group refractive index.

The angle between the wave normal and the ray direction can be represented graphically with the help of a polar plot in which the length of a radius vector drawn from an origin in the direction of the wave normal is equal to the refractive index μ. It can be seen from the construction of fig. 8.8 that the angle between the radius vector and the normal to the curve is then the same as the angle α between the wave normal and the ray direction. It is to be noticed that if $\partial\mu/\partial\theta$ is positive the angle is negative, in agreement with the sign in (8.117).

The preceding discussion was restricted to the situation where the amplitude in the wave front was limited only in the x-direction, so that the wave packet had the shape of an infinitely long strip. If, instead, it is limited in both the x- and y-directions the angular spectrum is in three dimensions. A simple extension of the previous analysis then shows that the ray direction is along the normal to the three-dimensional $\mu(\theta, \phi)$ surface at the point where the wave-normal direction intersects it.

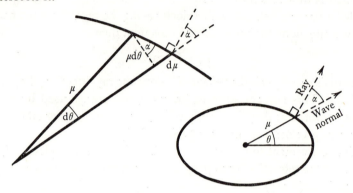

Fig. 8.8. The ray direction makes an angle α with the wave normal, such that $\tan \alpha = -\dfrac{1}{\mu} \dfrac{\partial \mu}{\partial \theta}$. α is thus the angle between the normal and the radius vector of a $\mu(\theta)$ curve in which μ is plotted as a function of θ.

9 Exploration of the ionosphere with radio waves†

9.1 Introduction

Experimental methods of two fundamentally different kinds have led to knowledge of the ionosphere. One makes use of equipment carried aloft by rockets or artificial satellites to study the ionosphere in their neighbourhood. The apparatus is usually a modification of that used in ground-based laboratory experiments; for example, useful results have been obtained with mass spectrometers, magnetometers, and Langmuir probes. The principles used are discussed in chapter 10. In experimental methods of the other type, use is made of radio waves emitted either by natural processes or by man-made transmitters. Methods of that kind were devised for exploration of the upper atmosphere and have not yet found much application in ground-based laboratories. This chapter is devoted to an account of the principles on which they are based: the results of the experiments are not discussed here, but at appropriate other places in the book.

The most important ways in which man-made radio waves have been used to study the ionosphere can be classified as follows:

(i) *Ionospheric sounding by total reflection.* Waves are reflected from the ionosphere, by a process similar to the total reflection of light waves, and the time interval between the emission of a wave and its return to the sending point is measured. The radio frequency of the emitted pulse is altered smoothly and the echo time is recorded as a function of frequency: the record is called an *ionogram*. The apparatus is called an *ionosonde*; when it is on the ground it is used to explore the lower part of the ionosphere, when it is in a space vehicle it is used to explore the upper part or topside; it is then called a *topside sounder*. The emissions from a topside sounder sometimes produce disturbances in the local ionosphere that last much longer than the emitted pulse; they have been called *resonances*, and have proved useful in ionospheric investigations.

† [13, 14, 16].

(ii) *Partial reflection*. The totally reflected waves recorded by iono-sondes are usually so strong that senders of comparatively small power can be used at places where electrical interference is not particularly small. If a high power sender is used in a place where there is little interference it is possible to investigate much weaker echoes returned after partial reflection from the ionosphere. In one method, the partial reflection occurs because the electrons have a distribution that is irregular on a scale much greater than the distance between them and much less than a radio wavelength; it is called the method of *partial reflection*. In another method the partial reflection represents the wave energy returned from individual electrons, each scattering inde-pendently, in the way first described by J. J. Thomson. It is called the method of *incoherent scatter* or of *Thomson scatter*.

(iii) *Wave interaction*. The temperature of the ionospheric electrons in a small defined region of the ionosphere is increased by the absorp-tion of energy from a short pulse of radio waves: the effect of this heated region on another wave traversing it is then observed.

(iv) *Doppler shift*. A wave emitted from a space vehicle is observed at the ground and the Doppler change of frequency, resulting from movement of the vehicle in the line of sight, or from changes in the intervening ionosphere, is measured.

(v) *Faraday rotation*. A linearly polarized wave emitted from a satel-lite has its plane of polarization rotated as it traverses the anisotropic ionosphere: the changes in the polarization of the wave received at the ground are observed as the satellite moves or as the ionosphere changes.

The most important naturally occurring waves that have been used in ionospheric research can be grouped as follows.

(vi) *Electron-whistlers* are audio-frequency electromagnetic waves radiated impulsively from lightning flashes near the earth; they are produced when the different Fourier components of the impulse travel through the ionosphere at different speeds so as to be received as a note of decreasing frequency, known as a *whistler*. Reception can be either on the earth, or in a space vehicle. Most whistlers provide evidence about the electron content of the regions they traverse. Certain types receivable in satellites and known as *electron-ion-whistlers*, also provide an estimate of the relative concentrations of ions of different masses in the neighbourhood of the satellite.

(vii) *Ion-whistlers* occur at infrasonic frequencies in the range 0.5 to 5 Hz. They have their origin in the action of the solar wind, or of streams of energetic particles, on the magnetosphere and they travel to the earth as hydromagnetic waves. Their spectrum at the receiver is determined by dispersion of these waves much as the spectrum of audio-frequency whistlers is determined by dispersion of electromagnetic waves.

(viii) *Very low frequency (VLF) noise* (sometimes called *hiss*) occurs naturally in the ionosphere. Although there is no accepted theory to account for its origin it is often found that its spectrum is terminated abruptly at a low frequency cut-off: by measuring this cut-off deductions have been made about the electron and ion content of the magnetosphere.

It is convenient to discuss the different experimental methods with reference to the dispersion curves for longitudinal and transverse propagation. For a situation where ions of two different masses are present these curves are sketched in fig. 9.1 and are labelled to show the frequencies of importance in the different experimental methods; the sketch is not to scale. Fig. 9.2 shows the magnitudes of some of the most important frequencies at different heights in the ionosphere.

9.2 Ionospheric sounding [31, 78]

As an introduction to the concept of vertical incidence sounding of the ionosphere, consider the behaviour of a wave that leaves the ground with wave normal making an angle i_0 with the vertical and neglect the influence of the geomagnetic field and of collisions. As the wave moves upwards it encounters an increasing concentration $n(h)$ of electrons, and hence a decreasing refractive index (μ) as given by

$$\mu^2 = 1 - (e^2/\epsilon_0 m\omega^2)\, n(h)$$

If the ionosphere is horizontally stratified, and if the wave normal at any height makes an angle i with the vertical, Snell's law requires that $\mu \sin i = \sin i_0$. If, at some height, the concentration of electrons is great enough to make $\mu = \sin i_0$ the wave normal becomes horizontal, the wave is reflected, and it returns to the ground: no energy penetrates above that height. The situation is analogous to that where a light wave falls obliquely on a sharp boundary separating a medium of greater refractive index from one of lesser: it is described as total reflection. If the angle i_0 is reduced to zero, so that the wave travels vertically, the condition for total reflection occurs at the height where $\mu = 0$ or where $n = (\epsilon_0 m/e^2)\,\omega^2$. If the electron concentration in an ionospheric layer

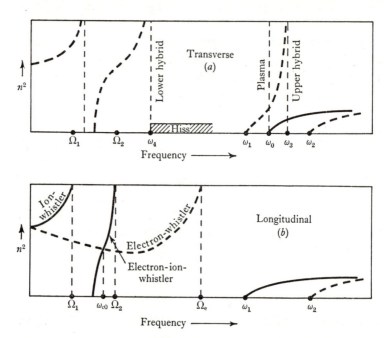

Fig. 9.1. Schematic dispersion curves (not to scale) for longitudinal and transverse propagation through an ionosphere containing electrons and two types of ions with gyro-frequencies Ω_e, Ω_1 and Ω_2. The parts of the curves associated with natural VLF radiations and with 'resonances' are marked.

has a peak value (n_m) at some height then frequencies less than $\omega_m = \sqrt{(n_m e^2/\epsilon_0 m)}$ will be reflected and those greater will penetrate. The penetration frequency ω_m is measured as a routine at numerous ionospheric observatories to provide information about n_m on a world-wide scale.

The presence of the geomagnetic field causes the emitted wave to travel as two separate characteristic waves, the penetration frequency of each is different, and the determination of the peak electron content is a little more complicated. Moreover in the presence of the magnetic field it cannot be supposed that the reflection point is necessarily immediately overhead. Thus, in a horizontally stratified ionosphere, it follows from Snell's law that if the wave normal is vertical when a wave leaves the ground it will remain vertical everywhere, but in an anisotropic ionosphere this does not imply that the *ray* direction is vertical. Suppose, for example, that with the geomagnetic field in the

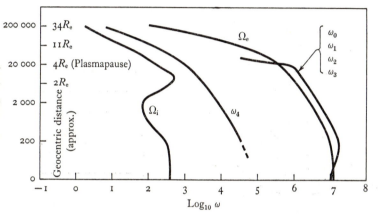

Fig. 9.2. The height variation of some of the important frequencies marked in fig. 9.1. Ω_i mean ion gyro-frequency; ω_4 lower hybrid frequency defined by (8.83); Ω_e electron gyro-frequency; ω_0 plasma frequency; ω_3 upper hybrid frequency defined by (8.82); ω_1, ω_2 defined by ((8.35) and (8.36) or (8.80) and (8.81)).

direction *AB* of fig. 9.3(*a*) the refractive index surfaces at increasing heights 1, 2, 3, 4, 5 in the ionosphere take the forms shown. Then at each level the ray direction (indicated by the arrows) is along the normal to the curve at the place where the vertical wave normal intersects it (§8.8). These ray directions are used, in fig. 9.3(*b*), to construct the path followed by a wave packet. Reflection occurs at a horizontal distance *d* from the overhead position which for a typical case may be as great as 30 km and is different for the two characteristic waves. Sometimes this difference is important.

In an ionogram, the echo time of a short burst of radio waves reflected at vertical incidence from the ionosphere is plotted against the radio wave frequency. From the curve it is possible to deduce the height-distribution of electrons in the following way. At any height (h) in the ionosphere a wave travels with a group refractive index $\mu'(h)$ related to the local electron concentration $n(h)$ in a known way: it is reflected at a height h_1 where the wave refractive index is reduced to zero. The travel time (t) of an echo is thus given by

$$t = (2/c)\int_0^{h_1} \mu'(h)\,\mathrm{d}h$$

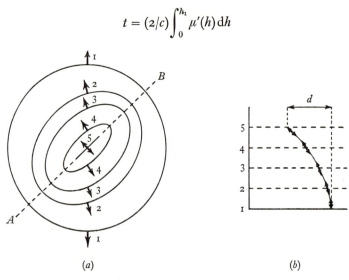

(a) (b)

Fig. 9.3. A wave packet leaving the ground with vertical wave normal encounters $\mu(\theta)$ curves like 1, 2, 3, 4, 5 in (a) at heights 1, 2, 3, 4, 5 in (b). The ray directions, being normals to the $\mu(\theta)$ curves, are shown by the arrows in (a). The corresponding path of the wave packet is shown in (b). At the height (5) of reflection it is a horizontal distance d from its original position.

and can be calculated for any distribution $n(h)$ and any frequency. Methods are available for performing the inverse calculation so as to deduce the height-distribution, $n(h)$, of the electron concentration from a knowledge of how the observed travel time t depends on the frequency of the radio wave. Examples of height-distributions calculated in this way are shown in fig. 9.4.

Collisions of electrons with heavier particles have been neglected in the preceding account. They are responsible for absorption, so that by observing the strength of the wave returned to the ground it is possible

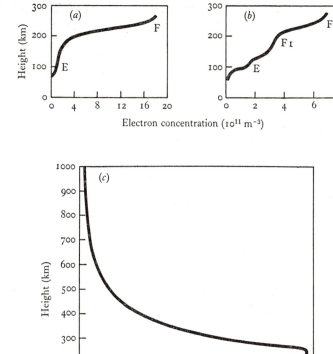

Fig. 9.4. Electron distributions deduced from ionograms. (*a*) From ground-based ionosonde when F 1 ledge was absent; (*b*) from ground-based ionosonde when F 1 ledge was present; (*c*) lower part from ground-based, and upper part from satellite-based ionograms. ((*a*) and (*b*) after reference 148, (*c*) after reference 56.)

to make deductions about the collision frequencies in the ionosphere. The collisions do not alter the real part of the refractive index sufficiently to invalidate the previous conclusions, except in one fundamental point that should be considered. In the presence of collisions the real part of the refractive index is nowhere reduced precisely to zero so that, according to the simple treatment outlined above, there

7

can be no total reflection of a wave incident vertically. To show why reflection occurs requires a 'full wave' theory, but the following simple treatment contains the essence of the matter.

Suppose that, over a limited portion of a medium where the mean refractive index is μ, there is a linear gradient of refractive index, and

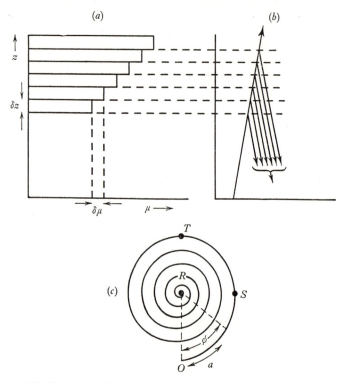

Fig. 9.5. To illustrate reflection from a medium in which the refractive index μ varies with height (after reference 26).

approximate, as in fig. 9.5(a), by supposing that it is made up of a series of equal infinitesimal steps, each of thickness δz, and that the refractive indices of adjacent steps differ by $\delta\mu$. Suppose also that a wave of unit amplitude is travelling vertically upwards. Then the simplest wave theory, of the type used in deriving Fresnel's reflection coefficients, shows that each boundary reflects a weak wavelet with amplitude $\delta\mu/2\mu$. Suppose that these wavelets are so weak that the original wave is effectively undiminished as it travels and that multiple reflections of the wavelets are negligible. Then, as indicated in fig. 9.5(*b*), the reflected wave is the resultant of all the partially reflected wavelets, when they have been added with due regard to their phases. If λ_0 is the free space wavelength the reflected wavelet is retarded in phase by $(2\pi/\lambda_0)(2\mu\,\delta z)$ at each successive step. The addition of these wavelets is represented on an amplitude-phase diagram by a regular polygon with sides of length $\delta\mu/2\mu$ having an angle $(2\pi/\lambda_0)(2\mu\,\delta z)$ between them. In the limit, when δz and $\delta\mu$ tend to zero, the polygon becomes a circle in which the relation between an element (da) of length and the element (dϕ) of angle subtended at the centre is given by

$$\mathrm{d}a/\mathrm{d}\phi = \lim\{\delta\mu/2\mu\} \div \{(2\pi/\lambda_0)(2\mu\,\delta z)\}$$

$$= (1/8\pi)(\lambda_0/\mu^2)(\mathrm{d}\mu/\mathrm{d}z) \tag{9.1}$$

If there is some absorption the main wave is weakened as it travels so that successive wavelets are weaker, and the circle becomes a spiral as in fig. 9.5(*c*). The resultant reflected wave is then represented by the radius OR which is equal to $(1/\pi)$ times the semi-circumference OST. But OST is given by

$$OST = \int_0^\pi (\mathrm{d}a/\mathrm{d}\phi)\,\mathrm{d}\phi$$

$$= (\lambda_0/8\mu^2)(\mathrm{d}\mu/\mathrm{d}z) \tag{9.2}$$

from (9.1). The amplitude of the resultant reflected wave is hence given by

$$OR = (\lambda_0/8\pi\mu^2)(\mathrm{d}\mu/\mathrm{d}z) \tag{9.3}$$

Equation (9.3), derived by admittedly crude reasoning, gives a correct account of what determines the magnitude of the wave reflected at normal incidence from a gradient of refractive index. In particular it shows that strong reflection occurs at a height where μ becomes small; that is why it is usually sufficiently accurate to neglect collisions and to assume that reflection is from the height where $\mu = 0$.

9.2.1 Resonances observed with topside sounders [46, 60, 164]

In a topside sounder, carrying a transmitter and a receiver, a short emitted pulse normally produces a correspondingly short response in the receiver at the time of its emission. On certain frequencies, however, the response lasts for a much longer time as though the ionosphere in the neighbourhood had been excited to resonate; the phenomenon is therefore called a *resonance*. It occurs when there are special relations between the frequency of the pulse and the nature of the surrounding ionosphere. With a variable-frequency sounder resonances are observed as the frequency passes through these special values; with a fixed-frequency sounder they occur when the sounder passes a part of the ionosphere that has the appropriate constitution. Resonances are observed at the plasma frequency, the electron gyro- (or cyclotron-) frequency, and the upper hybrid frequency; as marked on fig. 9.1.

The plasma resonance occurs when the frequency of the sender is nearly the same as the plasma frequency (ω_0) of the surrounding medium. Precisely at that frequency a wave with wave normal perpendicular to the magnetic field has zero refractive index and zero group velocity. On a nearby frequency the group travels with a finite but small velocity equal to the speed of the satellite: it is thus recorded as a long-enduring trace that lasts until it is attenuated down to the limit of detection. Unfortunately the plasma resonance provides only redundant information, since the plasma frequency can also be deduced from one of the echo traces on the ionogram.

The cyclotron resonance occurs when the frequency of the radiated pulse is equal to the gyro-frequency (Ω_e) of the electrons in the neighbourhood of the satellite. It provides a measure not of the electron or ion content but of the magnetic field.

The upper hybrid resonance occurs when the frequency (ω_3) of the sender is equal to $\{\omega_0^2 + \Omega_e^2\}^{\frac{1}{2}}$ [see (8.58): the small correction introduced by considering ions is usually negligible]. If the plasma frequency (ω_0) and the electron gyro-frequency (Ω_e) are known from other information a knowledge of the upper hybrid frequency does not add much. If, however, the electron concentration is small, as it is above the plasmapause (see fig. 9.2), some valuable observations can be made, as follows. Under these conditions $\omega_3 (\fallingdotseq \Omega_e + \frac{1}{2}\omega_0^2/\Omega_e)$ is not very different from Ω_e: it can then happen that long-lasting traces corresponding to cyclotron resonance (Ω_e) and to upper hybrid resonance (ω_3) overlap and beat at the difference frequency $\frac{1}{2}\omega_0^2/\Omega_e$. From measurements of these beats, and with knowledge of Ω_e, it has been possible to determine

ω_0 and to deduce very small electron concentrations, between 8×10^2 and $1.2 \times 10^5\,\mathrm{m^{-3}}$, in the region beyond the plasmapause [164].

9.3 Partial reflection [53, 90]

If the ionosphere contains irregularities a wave travelling through it and encountering a change of refractive index from μ to $\mu + \Delta\mu$ in a distance small compared with a wavelength is partially reflected with reflection coefficient $\Delta\mu/2\mu$. Although partial reflections of this kind are mostly very weak, they can be observed if a high-powered sender is used on an electrically quiet site. They have been used in the following way to measure the electron concentration in the lower ionosphere.

Short pulses of radio waves are emitted circularly polarized, first with left-handed sense (corresponding in the northern hemisphere to the ordinary wave of magneto-ionic theory) and immediately after-wards with right-handed sense (extraordinary wave), and are received after weak partial reflection at vertical incidence. The amplitudes E_x and E_o of the extraordinary and the ordinary echoes are measured and their ratio plotted as a function of height as shown in fig. 9.6. The wave frequency is chosen so that total reflection occurs at heights greater than those to be explored. It is assumed that the partial reflections are produced by gradients of refractive index and that the reflection coefficients are given by $R = \Delta\mu/2\mu$. The refractive indices for the two characteristic waves are known to have the forms

$$\mu_0 = f_0\{\omega, \Omega_e, \nu(h)\}\,n(h) \quad \text{and} \quad \mu_x = f_x\{\omega, \Omega_e, \nu(h)\}\,n(h)$$

where f_0 and f_x are known functions of the wave frequency (ω), the gyro-frequency (Ω_e), and the height-dependent collision frequency $\nu(h)$: $n(h)$ is the height-dependent electron concentration. The ratio of the two reflection coefficients ($\Delta\mu/2\mu$) that result from a sharp change (Δn) in n at any height is thus

$$\frac{R_x}{R_o} = \frac{f_x\{\omega, \Omega_e, \nu(h)\}}{f_0\{\omega, \Omega_e, \nu(h)\}} \tag{9.4}$$

Since it does not depend on the concentration n its height-variation can be calculated from an assumed height-variation of $\nu(h)$: it is shown as a dashed line in fig. 9.6.

The ratio E_x/E_o of the received electric fields is not, in general, the same as the ratio R_x/R_o of the reflection coefficients because the two waves suffer different amounts of absorption in passing up to the reflection level and back. The method of experiment consists in comparing the observed magnitude of E_x/E_o with the calculated magnitude of R_x/R_o to deduce the ratio a_x/a_o of the absorptions suffered by the two waves returned from a series of different heights.

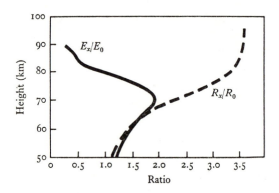

Fig. 9.6. Ionosphere sounding by partial reflection of waves from weak irregularities of electron concentration. The dashed curve represents the calculated ratios (R_x/R_o) of the reflection coefficients for the two characteristic waves reflected from weak irregularities at different heights: the height-variation of collision frequency is assumed known. The continuous curve represents the ratio (E_x/E_o) of the measured wave amplitudes partially reflected to the ground from different heights. The curves differ because, as the two characteristic waves travel between the ground and the reflection heights, they are absorbed by different amounts: from the ratio of the two absorptions the electron concentration at different heights is deduced (redrawn from reference 53).

The height-dependent absorption coefficients $k_0(h)$ and $k_x(h)$ for the two waves are known to be of the form

$$k_{o, x}(h) = g_{o, x}\{\omega, \Omega_e, \nu(h)\}\, n(h)$$

where the functions g are known. The absorption (a) of a wave reflected from a height h_1 is related to k by $a = \exp\left[-2\int_0^{h_1} k(h)\, dh\right]$

so that

$$a_x/a_o = \exp \left[\begin{array}{c} -2 \displaystyle\int_0^{h_1} g_x\{\omega, \Omega_e, \nu(h)\} n(h) \, \mathrm{d}h \\ +2 \displaystyle\int_0^{h_1} g_o\{\omega, \Omega_e, \nu(h)\} n(h) \, \mathrm{d}h \end{array} \right] \tag{9.5}$$

If then $a_x/a_o = (R_x/R_o)/(E_x/E_o)$ is determined, from curves like those of fig. 9.6, for a series of heights h_1 (9.5) can be used to determine $n(h)$. Electron concentrations at heights between 50 and 90 km have been obtained in this way. In fig. 9.6 it is noticeable that $R_x/R_o = E_x/E_o$ at the lowest heights, because there the absorptions are negligible.

9.4   Incoherent scatter or Thomson scatter [73, 80–2, 153, 171]

A wave falling on a free electron causes it to oscillate and to radiate a scattered wave in all directions. J. J. Thomson showed that the power re-radiated in this way was equal to the power in the incident wave that would fall on an area $\sigma = \pi\mu_0^2 e^4/m^2$, μ_0 being the permeability of free space: he called this area the scattering cross-section of the electron. Suppose that a wave carrying power P per unit area falls at a height h on a volume V in which the electron concentration is n. If the electrons are randomly distributed the lengths of the paths to each from the ground and back again are also randomly distributed, so that the scattered waves add with random phases: the powers then add and the total power returned to unit area on the ground is equal to $P\sigma nV/4\pi h^2$. If this returned power is measured n may be determined. The method of experimentation is called *incoherent*, or *Thomson scattering*. The volume V is determined by the geometry of the experiment. Sometimes pulses are used at vertical incidence and V is known from the pulse length and the sender or receiver beam width whichever is the smaller. Sometimes waves are set up continuously and observed from a distance with a narrow-beamed receiver: the volume V is then determined by the overlap of the sending and the receiving beams.

Because the Thomson scattering cross-section of an electron is so small, about 10^{-28} m², the power returned from the ionosphere is very small: for example, if the scattering volume has the dimensions of a cube with side 100 km at a height of 1000 km where the electron concentration is about $10^9\,\mathrm{m^{-3}}$, the total scattering cross-section is only $10^{-4}\,\mathrm{m^2}$. It is the power returned from this effective area, no larger than a coin, at a height of several hundred kilometers, that has to be measured. The measurements involve the use of very large antennae (the largest has an area of $3 \times 10^5\,\mathrm{m^2}$), working at places where the electrical noise is very weak, and it is necessary to use techniques of integration over times of the order of minutes.

The electrons that scatter the waves move with velocities that depend on their temperature, so that a wave emitted on a single frequency is returned as a band of Doppler-shifted frequencies whose spread depends on the velocities of the electrons in the vertical direction. At first sight it would seem that the frequency spreading should correspond to the thermal velocities of free electrons so that if it were measured the electron temperature could be deduced. For example, if a wave of frequency f were incoherently scattered from an ionosphere at $1000\,^\circ\mathrm{K}$, with electrons having r.m.s. speeds about $3 \times 10^5\,\mathrm{m\,s^{-1}}$ its frequency might be expected to be spread over a frequency band of order $(v/c)f = 10^{-3}f$. The measured frequency spreading is, however, about one hundred times smaller. The reason is that the electrons are not free, but are coupled by electrostatic forces to the ions: the situation needs a detailed discussion which shows that the controlling factor is the velocity of the ions and not of the electrons.

The fundamentals of this discussion can be understood as follows. Suppose first that the ionosphere consists of a scattering continuum and that a radio wave of wave number k (wavelength $2\pi/k$) travels vertically through it. Suppose also that a thin layer of thickness dh, at a height h, scatters back to the ground a weak wave of amplitude $s(h)\,dh$, $s(h)$ being a height-dependent scattering coefficient. The total field in the wave returned to the ground from a slab between heights of h_1 and h_2 is then proportional to

$$S(2k) = \int_1^2 s(h) \exp\left(2ikh\right) dh; \tag{9.6}$$

it depends, of course, on the wave number k. Equation (9.6) shows that $S(2k)$ is the component of spatial frequency $2k$ in the spatial Fourier transform of the scattering function $s(h)$. This conclusion can be expressed in words as follows:

If, in the slab of medium considered, the scattering coefficient $s(h)$ has a Fourier component $S(K)$ that has spatial wave number K, then an incident electro-magnetic wave with wave number $k = \frac{1}{2}K$ will be scattered backwards from it with amplitude proportional to $S(K)$. It is as though the incident wave selects that Fourier component in $s(h)$ that has a wave number equal to twice its own wave number, or a wavelength equal to half its own wavelength.

If, instead of being a continuum, the ionospheric slab consists of stationary electrons randomly distributed, the Fourier components $S(K)$ have amplitudes independent of K so long as their wavelength $\Lambda \, (= 2\pi/K)$ is greater than the mean distance between electrons: all radio wavelengths λ are thus scattered equally strongly provided λ is appreciably greater than the interelectron distance.

In the ionosphere, however, the electrons are not stationary, but are moving in a way that is strongly controlled by the ions. The possible motions are those of electron–ion-acoustic waves (§ 8.7), and these waves, of different length (Λ) each travelling with its own velocity $V(\Lambda)$, constitute the Fourier components of the electron concentration. The radio wave returned to the ground, scattered by an upward travelling Fourier component wave, suffers a (radian) Doppler shift of magnitude

$$\Delta\omega = 2\pi \times 2V(\Lambda)/\lambda = 2\pi V(\Lambda)/\Lambda = \Omega(\Lambda)$$

where $\Omega(\Lambda)$ is the frequency of the electron–ion-acoustic wave of length Λ. The downward travelling Fourier component waves produce Doppler shifts of the same magnitude but of opposite sign.

If, in the angular spectrum of the original radio beam, there is a Fourier component corresponding to a radio wave travelling at an angle to the vertical it is returned by a Fourier component of the electron–ion-acoustic wave travelling in that same direction, and suffers the same Doppler shift as the vertical travelling wave.

To calculate the Doppler shift experienced by a radio wave of length λ it is thus necessary to determine the frequency (Ω) of an electron–ion-acoustic wave of length $\Lambda = \frac{1}{2}\lambda$. The discussion of § 8.7 shows that in general for any one wavelength Λ there are two waves in the ionosphere, one the electron-acoustic wave, that has frequency

$\Omega \fallingdotseq \omega_{0e}$ whatever the wavelength, and one, the ion-acoustic wave, that has velocity V given by

$$V = v_i \sqrt{(1 + \alpha)}$$

where v_i is the r.m.s. thermal velocity of the ions and α is the ratio T_e/T_i between the temperatures of the electrons and the ions.

If the ion-acoustic wave has length Λ its frequency is

$$\Omega \fallingdotseq 2\pi v_i \sqrt{(1 + \alpha)}/\Lambda$$

and, with $\Lambda = \tfrac{1}{2}\lambda$, this expression gives the appropriate Doppler shift suffered by a radio wave of length λ.

An incident wave of length λ thus produces a scattered wave in whose spectrum there are two components that have suffered different Doppler shifts; each component is a pair with equal positive and negative shifts corresponding to the downward and upward travelling plasma waves. One is shifted by the plasma frequency, of order $10^6\,s^{-1}$, it is called the plasma line. The other, which might be called the ion-dominated line, is shifted by $\Omega = 2\pi v_i(1 + T_e/T_i)^{\frac{1}{2}}/\tfrac{1}{2}\lambda$ or, with

$$v_1 = (kT_i/m_i)^{\frac{1}{2}} \text{ and } T_e/T_i = 1, \text{ by } \Omega = \frac{2\pi}{\lambda}\left(\frac{8kT_i}{m_i}\right)^{\frac{1}{2}}$$

For temperatures of order 1000 °K, and a radio wavelength of 1 metre (300 MHz) this Doppler shift is about $10^4\,s^{-1}$, much smaller than the shift of the plasma line. The position of this line depends on the ion temperature through the factor $v_i \propto T_i^{\frac{1}{2}}$ but, as T_e/T_i becomes greater than unity it becomes progressively more nearly proportional to $T_e^{\frac{1}{2}}$ and thus dependent on the electron temperature.

The two Doppler-shifted lines have very different shapes. The electron-acoustic wave has a velocity much greater than the thermal velocity of the electrons or the ions, it suffers little Landau damping, has a sharply defined frequency, and produces a sharp line in the spectrum of the scattered radio wave. The ion-dominated wave, however, travels with a velocity that is near the r.m.s. thermal velocity of the ions, the Landau damping is therefore very great and the corresponding line in the spectrum of the scattered wave is broadened. The broadening is so great that it extends to zero frequency, and in the spectrum there are seen not two lines, with frequencies equally separated from the incident frequency, but one broad line, centred on the original frequency and spread on each side to about where the two separate lines might be expected.

If the ratio (T_e/T_i) of temperatures is greater than unity the Doppler shift of the ion-dominated line is greater than before: the difference between the velocity of the corresponding ion-acoustic wave and the r.m.s. velocity of the ions also increases, so that the Landau damping decreases and the shifted line is sharper. These phenomena are

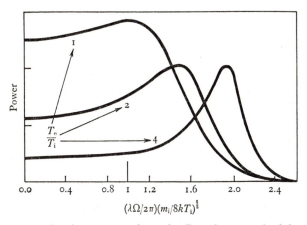

Fig. 9.7. Calculated curves to show the Doppler spread of frequency in waves returned from the ionosphere by incoherent scatter. A wave with a single frequency corresponding to zero on the frequency scale is returned as a band of frequencies. The figure shows one half of the symmetrically spread power spectrum appropriate to different ratios of electron- to ion-temperature. The frequency scale is normalized: λ is the original wavelength, Ω is the (radian) Doppler shift, m_i and T_i are the mass and temperature of the ionospheric ions (after reference 79).

illustrated in fig. 9.7 which shows that, when $T_e/T_i = 1$ the positively and negatively Doppler-shifted lines merge with each other to produce a broad line, whereas when $T_e/T_i = 4$ they appear quite separate and are shifted further from the incident frequency.

When a pulsed transmitter is used in a Thomson scatter experiment the scattered pulse, returned from a wide range of heights, is much longer than the transmitted pulse. A portion of it, equal in length to the transmitted pulse, is selected by a gate centred on some particular height and the power contained in it is measured, if necessary after being integrated for a sufficiently long time. The gate is moved in steps, and the measurement repeated, to determine the height-distribution of the scattered power. In principle a measurement of the power scattered from a particular range of heights should provide a measure of the electron concentration there, but, since absolute calibration of the equipment is difficult, it is more usual to deduce only the shape of the curve showing the height-distribution of the electrons. If absolute values are required the curve is calibrated by using a ground-based ionosonde to measure the concentration at the peak of the layer.

The spectrum of the wave scattered from any particular height is used to determine T_i (from the width of the Doppler-broadened line) and T_e/T_i (from the shape of the line).

The electron-acoustic wave that produces a Doppler shift equal to the plasma frequency is usually so weak that the corresponding line in the spectrum cannot be observed. During the daytime, however, the electron-acoustic wave sometimes undergoes Landau enhancement by photo-electrons that travel nearly with its own speed (see §8.7.1) and then the line can be observed and used in the following way. The returned pulse is passed through a narrow filter adjusted to select a frequency removed from the emitted frequency by an amount ω_1 that corresponds to the expected plasma frequency in some part of the ionosphere. If ω_1 has been chosen suitably energy will be scattered from those heights where the plasma frequency is equal to ω_1 so that the trace, consisting largely of noise, will be as shown at (a) in fig. 9.8. The filter is then set, in succession, to frequencies $\omega_2, \omega_3, \omega_4, \ldots$ and the experiment repeated to give traces like $b, c, d, \ldots$. If these traces are placed at positions along the horizontal axis that correspond to the frequencies $\omega_1, \omega_2, \omega_3, \ldots$ the curve drawn through the plasma line responses can be made to provide a plot of the plasma frequency as a function of height. It can be used, if desired, to calibrate and to check the height-distribution derived from measurement of scattered power.

9.5   Wave interaction [47, 85]

When a short train of radio waves passes a group of electrons in the ionosphere it heats them up: after it has passed they cool down again.

While they are hotter than normal the frequency of their collisions with heavy particles is increased so that if another wave passes through them it is absorbed more strongly than usual. The phenomenon is called *wave interaction*, or *ionospheric cross-modulation*: it can be used in the following way to determine the electron concentration at different heights.

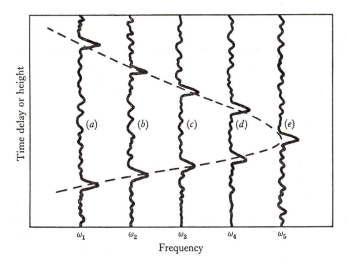

Fig. 9.8. Use of the 'plasma line' in incoherent scatter experiments. When the returned echo is passed through a filter that selects a frequency Doppler shifted by ω_1 it has a time-variation as shown at (a): it is enhanced at times corresponding to heights where the plasma frequency is equal to ω_1. The corresponding time variations when the filter is set to select other frequencies ω_2, ω_3, ..., are sketched in (b), (c), ...: from them it is possible to reconstruct the height-variation of the plasma frequency in the ionosphere.

A series of 'heating' pulses, each consisting of a short train of waves, is emitted at regular intervals together with a series of 'sounding' pulses emitted twice as often on a different frequency. The relative timing of the two sets of pulses is adjusted so that a sounding pulse travelling downwards after reflection in the ionosphere passes a heating pulse on its upward journey at some predetermined height h: as shown on the height–time diagram of fig. 9.9. Alternate sounding pulses (S) numbered 1, 3, 5, etc. arrive at the ground weaker than the pulses 2, 4, 6, etc. because they have passed through a layer of hotter electrons near height h. The amount of their weakening is measured: it depends on the energy absorbed from the heating pulse and on the absorption coefficient of the sounding wave. Both these quantities depend in a known way on the product $n\nu$ of the electron concentration n and the collision frequency ν. The magnitude of n at the crossing height is then deduced, the magnitude of ν at that height being assumed known. The crossing height (h) is varied, simply by altering the timing of the pulses, and the height-distribution $n(h)$ of the concentration is

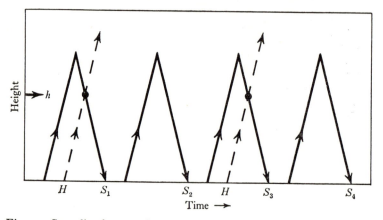

Fig. 9.9. Sounding by wave interaction. A pulse of radio waves (H) heats the electrons through which it passes. A sounding wave S passes the electrons immediately after they have been heated at a height h and suffers excess absorption. The heating wave is arranged to affect alternate sounding pulses: S_1, S_3, ... which are thus weaker than S_2, S_4, ...

deduced. The change in the strength of the sounding wave is small and the experiment requires a high powered 'heating' transmitter and a quiet receiving site.

9.6   Doppler and Faraday measurements [91]

If a source moving through the ionosphere with velocity V_1 emits waves of frequency f that travel with velocity V_0 in the direction of its motion then f waves are contained in a length $V_0 - V_1$ after unit time, so that their wavelength is $(V_0 - V_1)/f$. The frequency (V_0/λ) of these waves is hence $f(1 - V_1/V_0)^{-1} \fallingdotseq f(1 + V_1/V_0)$. If f and V_1 are known, V_0

can be deduced, and from it the refractive index, and the electron concentration, in the ionosphere surrounding the source can be calculated. In experiments that make use of this principle a rocket emits waves on two different frequencies; the first (f_1) is not far from the plasma frequency of the surroundings, so that the wave velocity (V_0) is different from the free space velocity, but the second (f_2) is so great that the wave velocity is effectively that of free space (c). Observation of the Doppler shift ($V_1/c)f_2$ of the second frequency gives a measure of the rocket's velocity (V_1) in the line of sight: observation of the Doppler shift, ($V_1/V_0)f_1$, of the first frequency then gives a measure of the wave velocity V_0 and hence of the electron concentration in the ionosphere surrounding the rocket.

The two characteristic waves that travel in the anisotropic ionosphere introduce complications. Their presence also leads to a phenomenon analogous to the Faraday rotation of the plane of polarization of a light beam when it travels through a transparent substance along the direction of a magnetic field. The phenomenon has been used in the following way to investigate the electron concentration of the ionosphere.

A satellite is caused to emit a linearly polarized radio wave on a frequency not too far removed from the ionospheric plasma frequency. In its travel through the anisotropic ionosphere the wave is split into two, which, under most circumstances, are effectively the circularly polarized characteristic waves appropriate to longitudinal propagation along the geomagnetic field. Each wave travels with its own proper velocity and when it arrives at the ground it combines with the other characteristic wave to re-form a linearly polarized wave whose plane of polarization differs from that of the original wave. Thus suppose that the two characteristic waves have refractive indices $\mu_o(s)$ and $\mu_x(s)$ at distances (s) along a path in the ionosphere, then the phase difference $\Delta\phi$ between them after travelling a distance s_1 is

$$\Delta\phi = (\omega/c)\int_0^{s_1} [\mu_o(s) - \mu_x(s)]\,\mathrm{d}s$$

if $\Delta\phi = 2p\pi$ the plane of polarization is rotated p times between the satellite and the ground. It is not usually possible to determine the total number (p) of rotations of the plane, but as the satellite moves from place A to place B the path of the wave changes and the plane of the received polarization rotates Δp times, where

$$2\pi\Delta p = [\Delta\phi]_{\text{path }A} - [\Delta\phi]_{\text{path }B}.$$

Each complete rotation of the plane through π is marked by a cycle of oscillation of the e.m.f. induced in a linear antenna: by counting these cycles, Δp can be determined as the satellite moves from A to B and as the path of the waves changes accordingly. If reasonable assumptions are made it is then possible to make deductions about that part of the ionosphere that lies between A and B. If the wave travels in a direction making an angle θ with the earth's field it is sufficiently accurate to suppose that the refractive indices μ_o and μ_x have values appropriate to longitudinal propagation in a field equal to $\cos\theta$ times the actual field.

Some modifications of the simple experiment have provided more detailed results. In one the sender is mounted on a geostationary satellite so that the path of the wave changes only a little: rotation of the received plane of polarization then gives information about changes in the total electron content in a column of unit cross-section between ground and satellite. In another modification observations are made when the satellite passes a position where the waves travel to the ground in a direction that is perpendicular to the geomagnetic field. Under those conditions the plane of polarization is not rotated, so that the total number of rotations observed at the ground as the satellite moves to some other position provides a measure of the number of rotations of the plane of polarization in the new position.

9.7 Whistlers and micropulsations [18, 21, 63, 64, 65, 101, 165, 166]

Because very large radiating elements are available in nature, electromagnetic waves of very great length (very small frequencies) can be radiated naturally with considerable intensity. Lightning flashes are long enough to emit impulses of radiation with component frequencies spread over a band from about 500 to 10^4 Hz: they are detectable with ordinary radio or audio amplifiers as *atmospherics* and *whistlers* of the kind discussed below. Waves of even greater length are also radiated

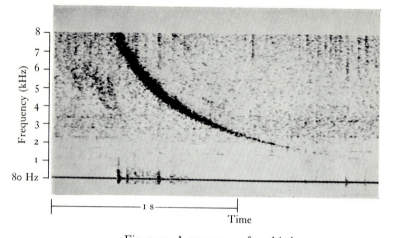

Fig. 9.10. A sonogram of a whistler.

By courtesy of Dr D. Jones, Cavendish Laboratory, Cambridge.

(*facing p. 199*)

from large volumes in the magnetosphere by the action of the solar wind sweeping over it, or by rapidly moving particles passing through it. The waves emitted have frequencies from a few kHz down to a few mHz (*milli*-Hertz not Mega-Hertz) with periods of about 10^{-3} s up to 1000 s. Those of frequencies greater than a few 100 Hz are receivable with the help of ordinary radio or audio frequency amplifiers, they are named *VLF emissions*: in telephones they produce sounds described as *hiss* or *chorus*. Waves of frequency less than 5 Hz are usually recorded through the action of their magnetic field and are called *geomagnetic micropulsations*.

9.7.1 Method of recording

The field variations associated with naturally occurring waves are usually complicated: sometimes they are quasi-periodic. It has proved convenient to examine their time-varying Fourier transforms by first recording the magnitude of the varying electric or magnetic field on a magnetic tape. If the modulation is in the range of audio frequencies the tape is then played back through a series of electrical filters whose time-varying outputs are exhibited on a photographic film in such a way that the blackening along any vertical line represents the strengths of the Fourier components of the field at one time: successive vertical lines correspond to successive times and indicate the change in the Fourier analysis as time progresses. The device is called a sonograph. When it is used to analyse a whistler the record has a shape like that of fig. 9.10. If the changing field has Fourier components with frequencies too great for the circuits of the sonograph the tape is slowed down before the analysis is made. If the changes are too slow (as in micropulsations) the tape is speeded up; sometimes it is speeded up and the modulation is re-recorded, the new tape is then itself speeded up and the modulation is again re-recorded and the analysis made on the final record.

9.7.2 Electron whistlers

An easily recognizable type of naturally occurring VLF radiation has a (roughly) single frequency that varies with time as indicated in the changing spectrum of fig. 9.10. When the lower (audible) frequencies are heard on a telephone receiver they produce a whistle of decreasing pitch. This type of natural radio wave is called a *whistler* and is produced when the impulsive radiation from a lightning flash travels through the ionosphere.

The frequencies (ω) to be discussed are between about 100 and 10000 Hz, in and just beyond the audible range: they are much smaller than the electron gyro-frequency (Ω_e). The wave thus travels in the mode marked electron whistler on the dispersion curve of fig. 9.1: the refractive index is greater than unity so that the wave penetrates the ionosphere without being reflected. After it has entered the ionosphere it is guided, for reasons to be discussed shortly, along a geomagnetic line of force: it travels far out into the magnetosphere and finally returns to earth at the magnetic conjugate point, where it can be recorded at the ground. The component frequencies in the original impulse travel at different speeds and arrive at different times as shown in fig. 9.10.

Sometimes, after a whistler has arrived at the far end of a line of force, it is reflected at the earth's surface and returns, doubly dispersed, to the place where it started. It may sometimes be reflected at both ends of the line so that it bounces back and forth many times to form a train of whistlers, each dispersed more than the preceding one.

Two different mechanisms assist in guiding the wave along the geomagnetic field. The first arises, as follows, from the anisotropy of the ionosphere. If the refractive index (μ) is represented in a polar plot as a function of the wave-normal direction, a curve like that of fig. 9.11 results. The normal to this curve, at the point where any wave-normal direction intersects it, represents the direction of travel of a wave packet (the ray direction): it is noticeable that whatever the wave-normal direction the ray direction can never depart from the field direction by more than about 20 degrees. The wave packet from the lightning stroke thus travels roughly along a geomagnetic field line up into the magnetosphere and back to earth at the conjugate point.

There is a second influence that guides the waves along the magnetic lines. Electrons in the ionosphere and magnetosphere are not uniformly distributed, but in many places are more (or sometimes less) concentrated along comparatively narrow tubes that follow geomagnetic field lines. Concentrations or deficiences are distributed in that way because diffusion occurs most readily along the field lines.

A tube-like irregularity of that kind constitutes a duct in which the wave, already tending to follow a field line, is trapped as in a wave guide. The different frequencies travel in the duct each with its own wave packet velocity and arrive in succession at the ground to produce the whistler. If the original pulse is also received without dispersion after it has travelled beneath the ionosphere it is possible to determine the time of travel of each component frequency in the whistler.

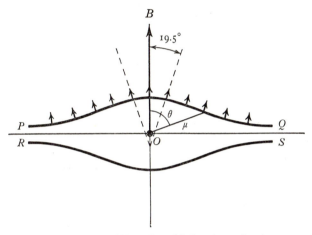

Fig. 9.11. To illustrate the guiding of a whistler along the geomagnetic field. The lines PQ and RS form a polar plot of refractive index (μ) as a function of the angle (θ) between the imposed magnetic field B and the wave normal of a wave travelling in the mode labelled 'electron-whistler' in fig. 9.1. If the diagram is rotated about the line OB, PQ and RS trace out a 3-dimensional polar plot of μ. For different wave-normal directions the ray directions are normals to the surface at the corresponding points, as represented by the arrows: the ray directions nowhere depart from the direction of the field by more than 20° (after reference 3).

A knowledge of these times of travel leads in the following way to knowledge about the magnetosphere out to the greatest distance reached by the guiding duct.

If the ionosphere were homogeneous with plasma frequency ω_0 and if the imposed magnetic field were uniform corresponding to electron gyro-frequency Ω_e then, for propagation with wave normal along the field direction (8.27) shows that one characteristic wave has

$$\mu^2 = 1 - X/(1 - Y)$$

or

$$\mu^2 = 1 + \frac{\omega_0^2/\Omega_e^2}{(\omega/\Omega_e)(1 - \omega/\Omega_e)} \tag{9.7}$$

The frequencies to be discussed here are in the near audio range ($\omega < 10^4\,\mathrm{Hz}$) and in the ionosphere $\omega_0 \fallingdotseq \Omega_e \fallingdotseq 10^6\,\mathrm{Hz}$ so that the first term (unity) can be neglected compared with the second in (9.7). With this approximation the group refractive index (μ_g) is deduced from (9.7)

$$\mu_g = \frac{\omega_0/\Omega_e}{2(\omega_0/\Omega_e)^{\frac{1}{2}}(1 - \omega/\Omega_e)^{\frac{3}{2}}}$$

If an impulse containing all spectral frequencies were to travel a distance l through a medium with this refractive index the time of travel $t(\omega)$ of a component of frequency ω would be

$$t(\omega) = \frac{\mu_g l}{c} = \frac{l}{c}\left(\frac{\omega_0}{\Omega_e}\right)\frac{1}{2(\omega/\Omega_e)^{\frac{1}{2}}(1 - \omega/\Omega_e)^{\frac{3}{2}}} \tag{9.8}$$

and if it were plotted against (ω/Ω_e) the curve would be like that of fig. 9.12. It has been thought by some that this curve resembles the shape of a nose and the frequency of minimum time delay (t_n) is called the *nose frequency* (ω_n). From 9.8 it follows that

$$\omega_n = 0.25\Omega_e \quad \text{and} \quad t_n = \frac{8}{3\sqrt{3}}\left(\frac{l}{c}\right)\left(\frac{\omega_0}{\Omega_e}\right)$$

An observation of ω_n would thus provide a measure of Ω_e and, if l were known, ω_0 could be deduced from an observation of t_n.

When a whistler travels through the ionosphere the situation is more complicated; the wave packet is assumed to follow some field line along which Ω_e would vary in a known manner, but it is not known which is the appropriate line. Moreover, even if the line were determined the variation of ω_0 along it is not known, nor is the wave normal everywhere along the field direction. It is nevertheless possible to calculate the function $t(\omega)$ for any given field line and any assumed height-variation of ω_0 by integrating an expression that is like (9.8) except that wave packet velocity replaces group velocity. When this is done it is found that the form of the $t(\omega)$ curve is roughly the same for all reasonable height-variations of ω_0, so that the height-variation cannot be deduced from the curve. The magnitude of ω_n does, however, depend on the field line chosen and is related to the magnitude of the electron gyro-frequency (Ω_{el}) at the top of the line by the approximate expression

$$\omega_n \fallingdotseq 0.4\Omega_{el} \tag{9.9}$$

(instead of $\omega_n = 0.25\Omega_e$ for a uniform plasma in a uniform field). The calculations show that the minimum time delay t_n is related to ω_n in a way that depends only a little on the assumed height-variation of ω_0, but comparison with a large number of observed whistlers, each with different values of ω_n and t_n, shows that a model in which

$$\omega_0 = C\Omega_e \tag{9.10}$$

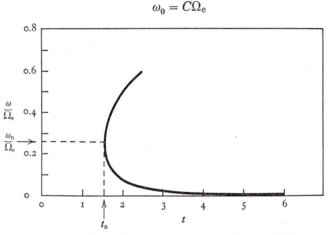

Fig. 9.12. The calculated frequency–time curve for a whistler travelling in a uniform ionosphere with plasma frequency ω_0 permeated by a uniform magnetic field. The frequency scale is normalized to the electron gyro-frequency Ω_e. The time scale, in seconds, represents the group time of travel over a distance $d(\Omega_e/\omega_0)$ where d is the distance travelled in unit time by a wave in free space (equal numerically to c). ω_n and t_n represent the frequency and the time-delay of the 'nose'. (After reference 18.)

fits the results best. Since the calculated relation between ω_n and t_n is not much dependent on the nature of the model it cannot be said that the experiments demonstrate the correctness of (9.10) but only that they are consistent with this supposition. The aim of most whistler researches has been to assume the validity of (9.10), to determine how the constant C changes from time to time, and to examine some special situations where (9.10) is no longer valid.

When a whistler is used to derive information about the ionosphere the procedure is thus as follows. The nose frequency (ω_n) is inserted in (9.9) to deduce the field strength (via Ω_{e1}) at the top of the field line, and then the time delay (t_n) is used to determine the constant C in (9.10).

The radiation from a single lightning stroke often travels for a long way beneath the ionosphere before it enters a duct to produce a whistler, so that sometimes a single stroke produces several whistlers travelling in different ducts and arriving in succession at the ground.† When multiple whistlers of this kind are analysed the magnitude of the constant C that fits all of them best is used. If the point of reception is at a moderate or low latitude nearby field lines reach only to comparatively small heights where Ω_{e1} is comparatively large: ω_n is then often too large to be recorded but it is possible to estimate its magnitude (and the magnitude of t_n) from the limited part of the $t(\omega)$ curve that is available.

In these ways it has been shown that over the equator and to geocentric distances of about 4 earth radii the magnetospheric electrons are distributed roughly in accord with (9.10) and that C varies with time of day, season and solar cycle. Sometimes whistlers from several successive lightning flashes follow one and the same duct: a particular duct being recognizable through the magnitude of ω_n (and hence Ω_{e1}) associated with it. A duct has occasionally been followed in this way for a long time (19 hours once): the important deduction can then be made that the duct, and presumably the magnetosphere in which it is embedded, rotates with the earth (compare p. 77). On other occasions, and particularly during a storm, ducts have been observed to move downwards at the equator, as though the geomagnetic field were being compressed (compare p. 93).

Whistlers observed in the polar regions travel on field lines that reach to comparatively great heights over the equator. They have led, in the following way, to a realization that the electron concentration over the equator obeys (9.10) with one value of C out to about 4 earth radii, but with a smaller value of C beyond that distance.

A single lightning flash in the polar regions sometimes excites several whistlers that travel in different ducts, some of which, like

† 'Multiple' whistlers of this kind are not to be confused with 'trains' of whistlers produced when a single wave packet travels repeatedly back and forth along a single field line.

1, 2, 3 in fig. 9.13 (*a*), are inside the plasmasphere while others such as 4, 5, 6 are outside. The nose frequency (ω_n) is smaller for the ducts that reach to the greater distances. In each of the regions separately the minimum time delay is greater when the duct path is longer (ω_n smaller), but the time delays for all those paths outside the plasmasphere are relatively less than for those inside, where the electron concentration is greater. The whistlers corresponding to the ducts 1 to 6 in (*a*) thus have shapes like those at 1 to 6 in (*b*). From records of this kind the position of the plasmapause can be measured and its changes followed.

9.7.3   Electron-ion whistlers [95]

When electron whistlers, caused by lightning, are observed in artificial satellites, a phenomenon attributable to the effect of ions is sometimes noticed. In the literature it has usually been called an ion-whistler but, to avoid confusion with the ion-whistlers discussed in the next section it is here called an electron-ion whistler. It can occur only when two (or more) species of ion are present. The dispersion curves for waves travelling along the magnetic field then take the forms shown in fig. 9.1 (*b*) and at the cross-over frequency labelled ω_{co} the refractive index is the same for the two characteristic waves. One of these, corresponding to the dotted curve, is the electron-cyclotron wave, or whistler mode: the other can be called an ion-cyclotron wave: it can travel at frequencies within the pass-band that is bounded at its upper end by the gyro-frequency Ω_2 of one of the ions. At the frequency where the curves cross energy can be transferred from one characteristic wave to the other (p. 155), so that a whistler travelling as an electron-cyclotron wave will excite an ion-cyclotron wave if it has the appropriate frequency.

The cross-over frequency is given by (8.47) and (8.48); it does not depend on the absolute concentrations of electrons or ions, but on the relative concentrations and masses of the two ions. Suppose that helium and oxygen are the important ions in the upper ionosphere and that the ratio of their concentrations is height-dependent so that the cross-over frequency varies with height as represented schematically in fig. 9.14(*a*), and suppose also that the upper limiting frequency (Ω_2) of the pass-band for the ion-cyclotron wave varies as shown,

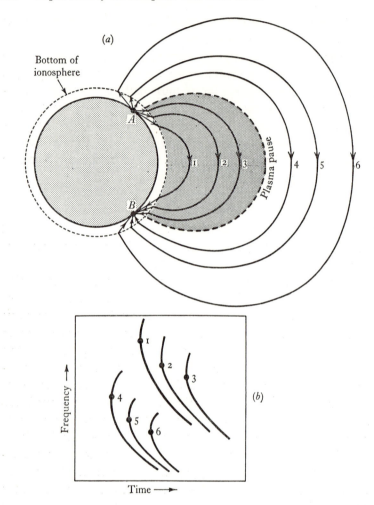

Fig. 9.13. A lightning discharge at A produces whistlers guided to B along ducts 1, 2, 3, 4, 5, 6 some of which are inside and some outside the plasmasphere. The received whistlers produce frequency–time traces as shown in (b). The nose frequencies decrease in the order 1 to 6 as the gyro-frequency (f_H) at the highest part of the field line decreases. The nose delay (t_n) increases as the length of the path increases: but it is relatively smaller for ducts 4, 5, 6 than for ducts 1, 2, 3 because the electron concentration outside the plasmasphere is smaller than that inside.

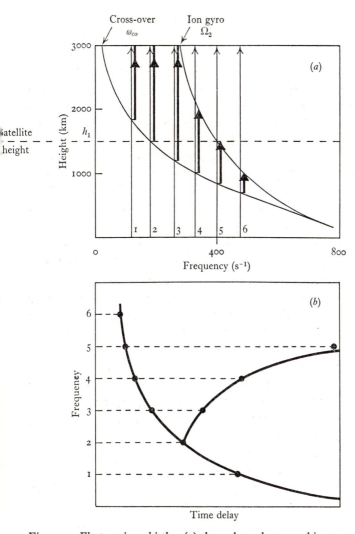

Fig. 9.14. *Electron-ion whistlers* (*a*) shows how the second ion gyro-frequency (Ω_2) and the cross-over frequency (ω_{co}) of fig. 9.1 vary with height. The thin vertical lines represent waves on frequencies 1 to 6 produced by a lightning discharge and travelling upwards in the electron-whistler mode: they excite additional waves (thick vertical lines) in the ion-whistler mode at the cross-over levels. The ion-whistler waves (thick lines) travel more slowly than the electron-whistler waves (thin lines) and cannot penetrate above the level where their frequency is the same as Ω_2. (*b*) shows the frequency–time trace that would be recorded in a satellite at the height shown in (*a*). The frequencies 1 to 6 correspond to those in (*a*).

because the geomagnetic field decreases upwards. Now suppose that a lightning flash radiates Fourier components of an electron whistler at frequencies marked 1, 2, 3, 4, 5, 6. They travel upwards as electron-cyclotron waves until they reach levels where their own frequency equals the cross-over frequency: there each shares its energy with an ion-cyclotron wave and thereafter both travel upwards with their own velocities. The ion-cyclotron wave-packet travels the more slowly: it cannot travel at all beyond the level where its frequency is equal to the ion gyro-frequency (Ω_2). The situation is represented for the different frequencies by the vertical arrows on fig. 9.14(a).

If a satellite is at a height h_1 it receives only the electron-cyclotron waves on frequencies 1 and 6, but on the other frequencies it receives the ion-cyclotron wave in addition, much retarded because the corresponding group velocity is small. Frequency 2 is the smallest one on which two waves are received and 5 is the greatest. The corresponding $t(\omega)$ trace (fig. 9.14(b)) is thus multiple over the range between frequencies 2 and 5. From traces of this kind it has been possible to measure the relative abundance of the ions (from the cross-over frequency) and also the ion-gyro-frequency at the height of the satellite (from the cut-off frequency 5).

Sometimes a situation arises where three ions (O^+, He^+ and H^+) are important. Then there are two cross-over frequencies, a second electron-ion whistler is recorded, and the relative abundance of the three ions can be determined.

9.7.4 Ion whistlers (micropulsations) [165]

A particular kind of geomagnetic micropulsation (known as pc1) consists of bursts of oscillation of frequency about 1 Hz that repeat at intervals of about 1 or 2 minutes. When observations are made at geomagnetic conjugate points the repeating bursts of oscillation are found to occur out of step as though they were caused by a group of waves travelling back and forth along a field line like the wave packet in a train of electron whistlers. It is believed that they are waves corresponding to the portion of the dispersion curve labelled ion-whistlers in fig. 9.1. They have been called *ion-whistlers* to stress the similarity to the waves that correspond to the part of the curve labelled electron-whistlers. The disturbances are supposed to be excited by bunches of energetic charged particles travelling through the magnetosphere.

9.7.5 Very low frequency (VLF) hiss [18, 46]

The intensity of very low frequency (VLF) noise recorded in a satellite is greater than that recorded on the ground: it often has a wide spectrum with a sharp low-frequency cut-off, of order 5–10 kHz. It has been called *VLF hiss*. There is no generally accepted theory of its origin but it has been established, experimentally, that the cut-off frequency coincides with the lower hybrid frequency (ω_4) of the ionosphere in the neighbourhood of the satellite. Equation (8.92) shows that the mean mass of the ions in the plasma can be deduced from a knowledge of ω_4 if the electron gyro-frequency Ω_e and the local plasma frequency ω_0 are known. Observations of the hiss spectrum made on topside sounder satellites that provided independent measures of Ω_e and ω_0 have been used in this way to determine the mean mass of the ionospheric ions at different times and places.

10 Experiments in space vehicles. Some fundamental principles†

10.1 Introduction

This chapter outlines some of the physical principles that are used when equipment is carried in a space vehicle to investigate the ionosphere in its neighbourhood. Details of apparatus are not discussed; they may be found in the review articles or the original papers listed in the bibliography, those numbered [10] and [154] contain many useful references.

A space vehicle disturbs both the neutral atmosphere and the ionospheric plasma in which it is situated. To avoid contamination of the atmosphere by gases emitted from the apparatus the measuring equipment is often de-gassed and sealed up in a vacuum before launch, the seal being broken only when it has reached the desired height.

In the ionospheric plasma a space vehicle is usually surrounded by a positive-ion sheath of the kind to be discussed below. It gives rise to serious difficulties even in a geostationary satellite, but when a satellite is moving rapidly through the ionosphere the situation is more complicated. Satellites move at speeds of order $8 \times 10^3\,\mathrm{m\,s^{-1}}$, and rockets at speeds that vary from about $10^3\,\mathrm{m\,s^{-1}}$ when they are low down to zero at the top of their trajectories. Although these speeds are usually smaller than the thermal speeds (of order $10^5\,\mathrm{m\,s^{-1}}$) of ionospheric electrons they are often greater than the speeds ($500\,\mathrm{m\,s^{-1}}$) of ions. The satellite thus sweeps the ions out of its way and trails a wake behind it where the ion concentration is much less than normal. The concentration of electrons is also small in the wake because electrostatic forces make them follow the ions [139, 168].

Apparatus for measuring the electron or ion concentration or temperature should be mounted on a support that projects from the vehicle to a distance of about twice the vehicle's radius and observations should be made when the apparatus is outside the 'wake'. If possible the apparatus should make use of a spherical sensor.

† [10].

During the day strong XUV radiation from the sun causes a large photo-emission from surfaces on a space vehicle so that currents flow between conductors that are held at different potentials. Sometimes special precautions must be taken to prevent these currents interfering with the intended measurements.

When a potential difference is applied between a space vehicle and an electrode mounted upon it, the potentials of the two in relation to space potential (see § 10.2) depend on the work functions of the two surfaces. In the ionosphere these work functions often change with time, as the result of oxidation or abrasion, in ways that can be very troublesome. Attempts are made to annul these changes by using surfaces that have been suitably prepared.

10.2 Measurements of electron and ion concentrations and temperatures [57]

This section is concerned with apparatus designed to investigate those ionospheric ions and electrons that are most numerous and that have temperatures less than a few tens of thousands of degrees, i.e. energies less than about 1 eV. The much less numerous particles, with larger energies measured in kilo-electronvolts, are investigated by other methods to be discussed in § 10.3.

As an introduction to the study of the particles with the smaller energies, consider a plasma in which there are n ions and electrons per unit volume moving with Maxwell distributions of velocity around the mean values v_e and v_i. A plane situated anywhere in this plasma is crossed, from both sides, by $\frac{1}{4}nv_e$ electrons and by $\frac{1}{4}nv_i$ ions in unit time: there is no resultant flow of either electrons or ions across it. If an open-mesh grid of wires were placed in this plane and its potential (with reference to some distant large conductor) were altered, it would in general alter the flows of the particles: there would, however, be one potential at which the flow would be practically unaltered (any small alteration would be produced by the collection of charges by the wires of the grid). This potential is called the *space potential.*

If now the open grid is replaced by a solid conductor the arriving electrons and ions give up their charges to it so that a current $\frac{1}{4}(nev_i - nev_e)$ flows to each unit area; since $v_e > v_i$ it flows in the sense corresponding to the arrival of electrons. This current continues to flow if the potential of the conductor is maintained, by external means, at the space potential.

If, however, the conductor (which might be a space vehicle) is isolated in space the excess electron current charges it negatively until it acquires a potential, called the *floating potential* at which there is no total current to it. At that potential, the conductor collects the charge from all the incident positive ions but only from those electrons with

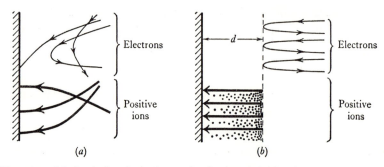

Fig. 10.1. (*a*) An isolated conductor in the ionospheric plasma acquires a potential more negative than the space potential. Most of the electrons are then repelled whereas all the positive ions are attracted, but because the electrons move more rapidly than the ions the total current is zero. (*b*) The situation can be idealized by supposing that there is a positive-ion sheath of width *d* outside the conductor from which all electrons are excluded and through which positive ions pass to the conductor as though through the space charge in a diode.

the greatest velocities in the Maxwell distribution; the ion current and the electron current are then equal and opposite. The conductor is then surrounded by a *positive-ion sheath* in which ions are more numerous than electrons: the situation is illustrated in fig. 10.1 (*a*).

Suppose that the electrons have temperature T, that the conductor is plane, and that its potential differs from space potential by V, just sufficient to reverse the velocity of an electron incident normally, so that $V = kT/2e$. Suppose also that all electrons have their velocities reversed when they reach a plane distant d from the conductor as shown in fig. 10.1 (*b*). The positive ions that cross that plane then reach the conductor through a space charge of other positive ions, as though they were travelling through a planar diode. The expression for the space-charge-limited current of density j_i through a diode then shows that

$$j_i = C\epsilon_0 \left(\frac{e}{m_i}\right)^{\frac{1}{2}} \frac{V^{\frac{3}{2}}}{d^2} \tag{10.1}$$

where C is a numerical constant of order unity and m_i is the mass of an ion. The ion current density across the 'diode' is the same at all distances and is equal to the current density in the plasma at the boundary plane so that

$$j_i = \tfrac{1}{4} n e v_i \tag{10.2}$$

If this value is inserted into (10.1), together with $V = kT/2e$ it is possible to obtain an expression for the thickness (d) of the positive-ion sheath: thus

$$d^2 = \frac{4C\epsilon_0 e^{\frac{1}{2}}(kT/2e)^{\frac{3}{2}}}{m_i^{\frac{1}{2}} n e v_i}$$

and with

$$m_i v_i^2 = 8kT/\pi \dagger$$

$$d^2 = (2\pi)^{\frac{1}{2}} C(\epsilon_0 kT/ne^2) \tag{10.3}$$

Thus, when the potential V is sufficient just to repel electrons incident normally with energy $\tfrac{1}{2}kT$, the sheath thickness is of order $(\epsilon_0 kT/ne^2)^{\frac{1}{2}}$, a quantity that plays an important part in the theory of plasmas, it is called the *Debye length*, and is usually denoted by λ_D. Table 7 lists the magnitude of this length in different parts of the ionosphere.

TABLE 7

Height h (km)	Debye length λ_D (m)
75	0.1
100	3×10^{-3}
150	3×10^{-3}
200	4×10^{-3}
400	4×10^{-3}
1200	4×10^{-2}
3000	6×10^{-2}
$10R_{\circ}$	2

If any piece of equipment is to be placed in a part of the ionosphere that is not affected by the sheath that surrounds the space vehicle it

† The root mean square velocity is $\sqrt{(3\pi/8)}$ times the mean velocity v_i.

must be held, on a support that projects from the body of the vehicle, at a distance greater than the sheath thickness.

10.2.1 Langmuir probes [57, 176]

In a device that has been widely used to investigate plasmas in the laboratory, a conducting probe, called a Langmuir probe, is inserted in the plasma and the current i flowing from it is plotted as a function of the potential difference V between it and earth. When the surface of the probe is plane the resulting curve has a shape like that of fig. 10.2.

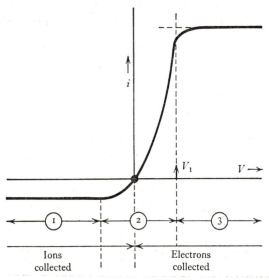

Fig. 10.2. To show how the current (i) flowing from a probe immersed in a plasma depends on the potential (V) applied to it. $V = 0$ corresponds to the floating potential at which $i = 0$: V_1 corresponds to space potential at which electrons and ions of all velocities reach the probe.

The part marked (1) corresponds to a situation where all the incident positive ions reach the probe but all the electrons are repelled; the part (3) corresponds to the complementary situation where no ions, but all the incident electrons, are collected. The part (2) corresponds to the situation where the electron current is carried only by the more energetic electrons in the Maxwell distribution. Because the total electron current is so much greater than the total ion current the shape of this part is largely determined by the electron current.

If u is the component of an electron's velocity perpendicular to the plane of the collector then, in a Maxwell distribution, the number with velocities between u and $u + du$ is proportional to $\exp(-mu^2/2kT)\,du$. If a retarding potential V is applied to the probe all electrons with $u > \sqrt{(2Ve/m)}$ can reach it, so that the number arriving in unit time, or the current (i), is proportional to

$$\int_{\sqrt{(2Ve/m)}}^{\infty} u \exp(-mu^2/2kT)\,du$$

thus

$$i = i_0 \exp(-eV/kT) \tag{10.4}$$

Here i_0 is the current when the retarding potential V is zero, i.e. when the collector is at the space potential V_1. That part of the curve marked (2) in fig. 10.2 is described by (10.4) if the (retarding) potential V is measured negatively from V_1. The object of an experiment is to find the shape of this curve and from it to determine the electron temperature T. Although the ion current, and various instrumental factors, distort the curve appreciably, methods are available to minimize their effects.

To transmit the i–V curve to the ground requires a considerable message-carrying capacity in the available telemetry; in one valuable technique the essential parameters of the curve are therefore determined in the satellite itself and are then transmitted to the ground with much less demand on the telemetry facitlities. Circuits are arranged that measure directly di/dV and d^2i/dV^2 to give

$$di/dV = (-e/kT)i_0 \exp(-eV/kT) \tag{10.5}$$

$$d^2i/dV^2 = (e^2/k^2T^2)i_0 \exp(-eV/kT) \tag{10.6}$$

Then $(di/dV)/(d^2i/dV^2) = -kT/e$ gives the temperature T and, when this is known i_0 can be calculated from the magnitude of di/dV at a known value of V. Knowledge of the probe's area A then allows the concentration n of electrons to be calculated from the relation

$$i_0/A = j_0 = \tfrac{1}{4}nev_e = \tfrac{1}{4}ne(8kT/\pi m)^{\frac{1}{2}} \tag{10.7}$$

The simple theory is appropriate to a Langmuir probe in a space

8

vehicle only because the velocities of the electrons are much greater than the velocity of the vehicle. If, however, an attempt is made in a satellite to repeat the experiment for positive ions, by applying a positive potential to the probe to repel the fastest ones, a different situation occurs because, for them, the satellite's velocity v_{sat} is greater than the gas-kinetic velocity. If the gas-kinetic velocity were negligible the positive ion current to a plane surface normal to the satellite's velocity would cease when the retarding potential V was such that

$$eV = \tfrac{1}{2}m_i v_{sat}^2 \qquad (10.8)$$

and if it were measured the mass m_i of the positive ions could be determined.

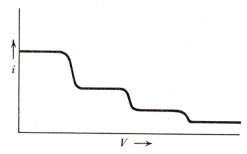

Fig. 10.3. When a satellite with velocity v_{sat} encounters ions with components of thermal velocity v_θ perpendicular to its surface the relative velocity is $v_{sat} \pm v_\theta$ with $v_{sat} > v_\theta$. When a variable retarding potential V is applied to a probe on the satellite the ion current (i) thus decreases when $V = \tfrac{1}{2}mv_{sat}^2/e$ and the shape of the $i(V)$ curve near this decrease depends on the distribution of the thermal velocities v_θ. If ions of different mass are present there are several steps in the $i(V)$ curve.

This method of experimenting has been used to show that the most important ions in the topside ionosphere are H^+, He^+, and O^+ and to determine their relative abundances. When the potential V is made successively more positive the current to the probe varies as shown in fig. 10.3: the potentials at which the current changes rapidly are substituted into (10.8) to give the ionic mass, the magnitudes of the corresponding changes in the current give measures of the relative abundances of the ions.

If the gas-kinetic velocities of the ions were negligible compared with the velocity of the satellite the steps in the curve of fig. 10.3 would be abrupt at the potentials given by (10.8). Because the gas-kinetic velocities are not zero the steps are, however, gradual and from their shapes the velocity distributions can be deduced: it is then possible to estimate the temperatures of the corresponding ions.

A probe mounted on a space vehicle must sometimes be brought to a potential more negative than the floating potential so as to draw

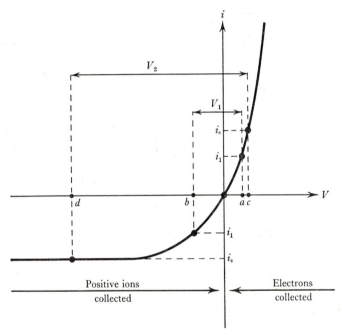

Fig. 10.4. To indicate what happens when a potential difference (V_1 or V_2) is applied between two conductors of equal area situated in the ionospheric plasma. The currents to the two must be equal and opposite.

positive ion current, and sometimes it must be more positive so as to collect electrons and negative ions. But it is often very difficult to apply a large positive potential to a probe particularly if it has an area comparable with that of the main vehicle. To understand the situation, suppose first that the probe and the vehicle have equal areas and that a potential difference V_1 is applied between them as indicated in fig. 10.4. The currents then differ from those that flow at the floating potential. The part of the system that is positive takes the potential a

and draws electron current i_1 while the part that is negative takes the potential b and draws positive ion current $-i_1$; these currents must be equal and opposite in a system that is isolated in space. Because the i–V curve is asymmetrical the potential difference V_1 is applied around the zero value (the floating potential) as shown. The application of the potential difference V_1 has thus caused the potentials of the two parts to differ from floating potential ($V = 0$) by different amounts; the potential of the electron collecting part has changed less than that of the ion collecting part. If the applied potential difference were so large

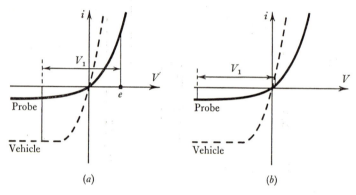

Fig. 10.5. To illustrate what happens when a potential difference V is applied between two conductors of unequal area situated in the ionospheric plasma. In (a) the larger (vehicle) is made more negative than the smaller (probe): in (b) the potential difference is reversed. In each situation the currents to the two conductors are equal and opposite.

that the saturation ion current (i_s) were drawn by one part of the system as represented by V_2, then the potential of the other part could never be greater than c, just sufficient to supply the same current of electrons.

   In practice the area A_v of the vehicle is greater than the area A_p of the probe and it is able to collect a correspondingly greater current: the situation is then as represented in fig. 10.5 where the i–V curve appropriate to the vehicle (shown dotted) has its current scale increased in the ratio A_v/A_p. If, then, the potential difference V_1 is applied to make the probe more positive than the vehicle the situation is as shown at (a). When the probe draws electron current equal to the saturation positive ion current that flows to the vehicle, the potential (e) is appreciably greater than the potential c of fig. 10.4 appropriate to a smaller vehicle. If it is required to draw the saturation electron current to the probe then the vehicle must be large enough to draw at least an equal and opposite ion current, so that

$$A_v nev_i \geqslant A_p nev_e$$

or
$$\frac{A_v}{A_p} \geqslant \frac{v_e}{v_i} = \sqrt{\frac{m_i}{m_e}} \fallingdotseq 170 \qquad (10.9)$$

If the vehicle is a satellite moving with a velocity v_{sat} greater than the velocity of positive ions then v_{sat} should replace v_i in (10.9). There is little difficulty in achieving the required ratio of areas for a Langmuir probe, but it is sometimes impossible to hold a larger piece of apparatus at a potential much more positive than the floating potential.

If the potential difference V_1 is applied so as to make the probe more negative than the vehicle the situation is as in fig. 10.5(*b*). There is no difficulty in making the probe much more negative than floating potential.

If the probe is small compared with the Debye length (λ_D) it collects electrons effectively from a sphere of area $4\pi\lambda^2_D$ and this is to be taken as the area A_p in (10.9). At heights less than 75 km where

$$4\pi\lambda_D^2 \fallingdotseq 10^{-1} m^2$$

it becomes increasingly difficult to make the probe collect any large electron current. This situation is particularly unfortunate in experiments with mass spectrometers used in the D region, where it is important to investigate the negative ions that accompany the electrons.

10.2.2 Radio frequency measurements in the ionospheric plasma [108]

The plasma surrounding a space vehicle can be investigated in different ways by observing its radio frequency behaviour.

In one type of experiment [116] the dielectric constant of the plasma is determined by using a radio frequency method to measure the capacity of a parallel plate condenser immersed in it. If there were no sheaths round the plates the dielectric constant would be equal to

$1 - ne^2/\epsilon_0 m\omega^2$. To remove the effect of sheaths the whole of the condenser is supported at a distance from the satellite, a sawtooth wave form is applied to it, and the radio frequency measurements are made when the potential corresponds to V_1 in fig. 10.2 so that there is no positive-ion sheath.

In other types of experiment an alternating potential of varying radio frequency is applied to a probe, which may be an antenna of considerable length, and the frequencies are noted at which various kinds of 'resonances' occur. One type of experiment makes use of the 'resonances' that are observed when a transmitter and a receiver carried by the craft are both tuned to certain frequencies; it has already been described in § 9.2.1.

Another type of 'resonance' is noticeable when the steady current to a probe is measured while a radio frequency potential is applied to it. There is an increased current when the frequency is related in a simple way to the plasma frequency of the surroundings: the phenomenon is called *resonance rectification*. Although the theory of the phenomenon is not properly understood, and although the 'resonance' does not occur precisely at the plasma frequency, some useful results have been obtained with apparatus that makes use of this phenomenon.

A more satisfactory experimental method [137, 138] depends on the observation that the complex impedance of an antenna exhibits sharp 'resonances' at two frequencies, one near the local plasma frequency, and the other precisely at the local value of the upper hybrid frequency. The resonance that occurs near the plasma frequency is influenced, to some extent, by the properties of the sheath around the antenna, and is not made use of; the other resonance occurs precisely at the hybrid frequency and is unaffected by the sheath: the reason for this is not clear. Measurements of this resonance have been used to determine the upper hybrid frequency $(\omega_{0e}^2 + \Omega_e^2)^{\frac{1}{2}}$ from which ω_{0e} is deduced if Ω_e is known.

10.3   Measurements on energetic particles

In addition to the ionospheric electrons and ions with energies up to about $1\,\mathrm{eV}$ there are electrons and ions with energies up to about $10^4\,\mathrm{eV}$ in the solar wind, up to about $10^8\,\mathrm{eV}$ in the zones of trapped radiation, and about 10^{10} eV in the solar protons that are emitted at times of a 'proton event'.† Experiments have been made in satellites to determine the energy distribution of these particles and how it depends on their direction of travel with respect to the magnetic field.

The simplest way of measuring the flux of energetic particles is to collect them in a Faraday cage and measure the resulting current. A grid in front of the cage must be held at a potential sufficient to exclude particles with one sign of charge, so that those with the other sign can be measured. Apparatus of that simple kind has been used, mainly in investigations of the solar wind.

It is more common to use equipment that will detect and count individual particles, and that will, in some cases, provide a measure of their energy. All types of detector depend on the production of a large number of secondary or tertiary charged particles when a primary particle of high energy enters the apparatus. The secondary particles may be produced in a gas, as in gas-filled counters, or at the surface of a solid, as in the channel multiplier: or in a scintillator where the primary particle produces a pulse of light which is then used to produce photo-electrons.

Tubular or spherical containers, of glass or thin metal, filled with gas into which an insulated electrode projects, have been used as gas-filled counters. The secondary ions, produced in the gas by an incident energetic particle, are attracted to the electrode; the gas pressure and the potential of the electrode determines the mode of action of the counter. If the potential is sufficiently small no tertiary ions are produced by the secondaries as they cross the counter; the charge (q) collected on the electrode is proportional to the number of secondary ions and hence to the energy of the primary particle. The charge is too small to measure, but if n primary particles enter the chamber in unit time, the resulting current, equal to nq, is measurable and is proportional to the flux of energy in the incident particles. A device working in that way is called an *ionization chamber*.

Sometimes the gas pressure and the potential are adjusted so that the secondary ions produce many tertiary ones as they travel to the electrode; the resulting pulse of charge is then great enough to be

† There are also galactic cosmic rays with energies up to about 10^{20} eV but they are not discussed in this book.

measured and is proportional to the number of secondary ions, and thus to the energy of the primary particle. A device of that kind is called a *proportional counter*. The pulses provided by the counter can be subjected to pulse-height analysis to reveal the energy distribution of the incident particles.

In another type of counter the gas pressure and the potential are such that each incident particle initiates a self-multiplying and self-sustaining avalanche of ionization. Arrangements are made to quench this discharge quickly so that a count of these (comparatively large) pulses provides a count of the particles incident even when they arrive in rapid succession. The size of the pulse is not however, proportional to the number of secondary ions and the device merely counts the particles regardless of their energy. It is called a *Geiger–Müller counter*.

In a different kind of proportional counter, called a *channel multiplier*, the secondary particles are produced, not by ionization of a gas, but by secondary emission from a solid surface. It consists of a thin circular tube coated inside with a material that emits secondary electrons easily when bombarded with primary particles of great energy; this material is also highly resistive and a potential difference, of order a few thousand volts, is applied between the ends of the tube so that there is a steep potential gradient down it. An energetic particle hitting the inside of the tube at one end produces secondary electrons that have sufficient energy to cross the tube; as they cross they are accelerated down the tube and when they hit the other side they produce tertiary electrons which, being themselves accelerated, produce still more electrons. The process is repeated several times. The number of electrons emerging at the far end of the tube is proportional to the number of secondary electrons emitted when the primary particle impinges on the tube; it is thus proportional to the energy of the incident particle.

In another device, which works in quite a different way, a high energy particle passes through a slice of transparent material and excites it to emit a pulse of light which is then converted into a pulse of current by a photo-multiplier tube. A device of that kind is called a *scintillator*; the output of the photo-tube is proportional to the energy that the incident particle deposits in the crystal. The scintillator must be covered with a film of opaque material (usually aluminium) thick enough to prevent sunlight's reaching it, or the photo-tube, but thin enough to transmit the particles to be counted.

When protons and electrons covering a wide range of energy are incident on a detector it is comparatively easy to determine the total number arriving in unit time and to make some estimate of their energy distribution, but auxilliary apparatus must often be employed to determine how their velocities are distributed in space, how many are protons, and how many are electrons. Direction of arrival is usually determined either by means of apertures in absorbing screens placed in front of a detector, or by pairs of detectors connected to a coincidence counter that records only when a particle has passed through them both.

Different absorbing materials, of different thickness, are sometimes placed in front of the detectors to distinguish between protons and electrons, or to remove all particles with energy less than some predetermined threshold. Sometimes electrons are excluded from a device intended to count protons by imposing a magnetic field just in front of the aperture. It is, however, difficult to use magnetic deflection without producing stray fields that interfere with measurements of the geomagnetic field that are often required on the same vehicle.

The numbers of particles in different energy ranges can be measured by passing them through two grids between which there is a retarding potential to stop the less energetic ones. It is, however, more common to use electric deflection of the particles in a gap between two concentric spherical electrodes, as shown in fig. 10.6. If an electric field E is directed across the gap, and if the particles enter the gap tangentially, those of energy U (non-relativistic) have the same radius (r) as the gap, and emerge at the far end if $U = \frac{1}{2}Eer$ while particles of other energies strike the walls. Particles of different energies are thus counted at the far end by altering the field E.

10.4 Mass spectrometers

Mass spectrometers have been used in space vehicles to determine the masses of ions, or of neutral particles, in the ionosphere: when neutral particles are to be investigated they are first ionized by electron bombardment. The ions, whether coming from the ionosphere direct, or after being ionized locally, are accelerated through a potential difference (V) and enter a mass spectrometer, which may have one or other of the forms described below.

Since the action of mass spectrometers depends on the free motion of ions in electric or magnetic fields, collisions with other particles must not interfere; the mean free path must thus be large compared with the dimensions of the equipment. Reference to Table 8 shows

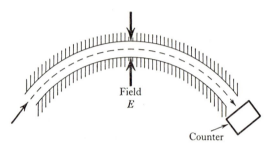

Fig. 10.6. Charged particles entering a gap between two concentric spherical conductors across which there is an electric field E will remain in the gap if $mv^2/r = Ee$.

that the free path of an electron or ion is large compared with the size of normal apparatus at heights above 100 km, but that below about 80 km (in the D region) too many collisions take place inside any apparatus of normal size. For experimenting at these heights it is thus necessary to reduce the pressure inside the spectrometer by continuous pumping and to cause the ions to enter through an aperture so small that the pressure difference between the outside and the inside can be maintained.

10.4.1   The magnetic deflection mass spectrometer

In one type of spectrometer a magnetic field B is applied perpendicular to the velocity v of the ion so that it follows a circular path of radius

$$r = \frac{mv}{eB} = \frac{1}{B}\left(\frac{2mV}{e}\right)^{\frac{1}{2}} \tag{10.10}$$

those following a selected circle, with one particular value of r, are collected. The accelerating potential V is swept back and forth so that ions with different masses move, in succession, along the selected path and are counted.

The need for a magnetic field, produced by a permanent magnet, makes this type of instrument rather heavy, and leads to possible interference with other measurements. These mass spectrometers have thus been used mostly on rockets and have generally been avoided on satellites where weight is more important and where stray magnetic fields are likely to be comparable with the earth's field.

TABLE 8. *The free path lengths of electrons and ions in the ionosphere*

Height (km)	Free path	
	of electrons (m)	of ions (m) $(O^+$ at 100 V)
50	1.7×10^{-3}	10^{-4}
60	5.3×10^{-3}	5×10^{-4}
70	1.9×10^{-2}	2×10^{-3}
80	8.3×10^{-2}	1.3×10^{-2}
90	6.4×10^{-1}	9.8×10^{-2}

10.4.2 The Bennett high frequency mass spectrometer [54]

In another kind of instrument, known as the Bennett mass spectrometer, the ions previously accelerated through a potential difference V pass in succession through three grids, separated by distances s, as shown at 2, 3, and 4 in fig. 10.7. An alternating potential is applied between the centre grid and the two outer grids which are connected together. Suppose first that this potential oscillates between $\pm V_1$ with $V_1 < V$, and that it has the square wave-form shown in fig. 10.7 so that it reverses at intervals τ. An ion, with charge of the appropriate sign, passing grid 2 just as the field reverses, finds itself in an accelerating field. If its velocity $v = s/\tau$ it reaches grid 3 just as the field reverses again, so that it is again accelerated between grids 3 and 4. In each portion of its travel, first from 2 to 3 and second from 3 and 4, it thus gains energy eV_1 and emerges with total energy $e(V + 2V_1)$. If it passes grid 2 at some other time it experiences a retarding field over part of its travel and passes grid 4 with a velocity less than $e(V + 2V_1)$: a particle with a different velocity, passing grid 2 at any time, also experiences a retarding field during part of its travel. The only ions that pass grid 4 with the maximum energy $e(V+2V_1)$ are thus some of those that have the correct velocity $v = s/\tau$. They are selected as they pass from grid 4 to grid 5 against a retarding potential difference that is slightly less than $(V+2V_1)$ and are counted on the far side of grid 5. Their velocity

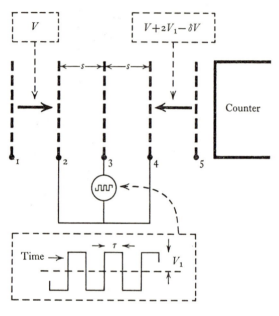

Fig. 10.7. The essentials of the Bennett high frequency mass spectrometer, described in detail in the text.

v $(= s/\tau)$ is related to their mass by $eV = \frac{1}{2}mv^2$. In practice the alternating potential difference is sinusoidal, not square-wave, and the detailed calculation is a little more complicated: but the principle remains the same. Spectrometers of this kind are simple and can be very light; they have been widely used.

10.4.3   The quadrupole mass spectrometer [144]

The quadrupole spectrometer consists of four parallel conducting cylinders placed at the corners of a square, opposite pairs are connected together, and a potential difference $V_0 + V_1 \cos \omega t$, is applied between them, part constant and part alternating. The ion beam is sent down the axis of symmetry of the system. Somewhat complicated theory shows that when the potentials are properly arranged most ions are deflected out of the beam and strike the conducting cylinders, but those with a certain mass, determined by the ratio V_1/V_0 of the alternating field to the constant field, can traverse the structure without being captured and can be counted at the far end.

10.5 Measurement of magnetic fields [123]

Magnetometers have been used in rockets to explore the magnetic field in the neighbourhood of the atmospheric dynamo (at 100–150 km), and in satellites to explore the field in the neighbourhood of the magnetopause. When the direction of the field is required it has been usual to use the 'flux-gate' principle; when only the scalar field is required devices depending on the Zeeman effect have been used.

10.5.1 The flux-gate magnetometer

The flux-gate magnetometer makes use of a bar of magnetic material that has a B–H curve of the form shown in fig. 10.8(a). To understand its action suppose first that the curve had the simple shape shown in (b). A sinusoidal magnetizing field $H \sin(2\pi n t)$ is applied so that the induction B has the shape of a 'gated' oscillation (hence the name flux-gate) as shown at (c). If the zero level is taken as shown in the figure, this oscillation consists of a train of square pulses repeating with a frequency n, and each of width $1/2n$. Its spectrum thus consists of sharp lines at frequencies that are multiples of n, enveloped by the Fourier transform of one of the pulses: it is shown in fig. 10.8(d). Because each rectangular pulse has width exactly equal to $1/2n$ its Fourier transform is zero at even multiples of n so that all the even harmonic lines are absent from the spectrum.

Suppose now that the small constant magnetizing field H_0 to be measured is superimposed on the sinusoidal one, as in fig. 10.8(e), so that each pulse lasts slightly longer than $1/2n$, its Fourier transform is then compressed slightly so that the spectrum of the pulse train is as shown at (g); even-order harmonics are now present. By measuring

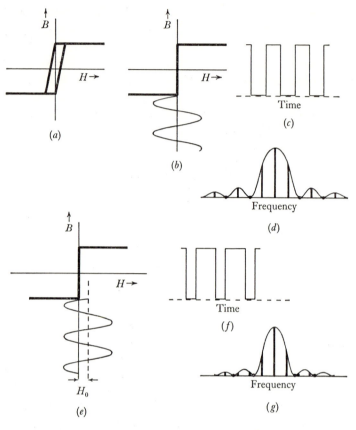

Fig. 10.8. To illustrate the action of a flux-gate magnetometer. The applied sinusoidal field (H) shown at (b) produces the changes of induction (B) shown at (c) and they have the spectrum shown at (d). The field applied at (e) produces the changes of induction shown at (f) and they have the spectrum shown at (g). The spectrum at (g) has even-order harmonics whereas that at (d) has not.

these the widening of the pulse, and thus the magnitude of the applied field H_0, can be determined. When the B–H curve has the more realistic shape shown in fig. 10.8(a) an argument of a similar kind can be used to show that the even harmonics appear only when there is a steady superimposed field. It can also be shown that the magnitude of the second harmonic is proportional to the applied field H_0.

H_0 is the field to be measured: the oscillating field is usually applied at a frequency of order 40 MHz. The odd-order harmonics are very much (about 10^6 times) larger than the even-order ones to be measured; they are removed to a large extent by using two magnetic rods (fig. 10.9) to which the alternating fields $H = \pm H \sin(2\pi nt)$ are applied in anti-phase while the field (H_0) to be measured acts in the same direction on both. A coil wound over both rods in the same direction is thus threaded by odd-order harmonic fluxes that cancel whereas even-order harmonics add: it is connected to a circuit tuned to the induced e.m.f. at frequency $2n$.

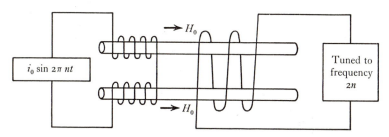

Fig. 10.9. A flux-gate magnetometer is frequently arranged with two ferro-magnetic rods in such a way that only the even harmonics induce e.m.f.s in the secondary coil.

In some versions of the experiment a null method is used in which the field H_0 to be measured is cancelled by an applied field: it is arranged that, if the fields become out of balance, the resulting output signal alters the applied field until balance is restored.

Flux-gate magnetometers measure the component of the field in the direction of the magnetic rods; they are sensitive down to about 10^{-9} T (0.1 gamma) and can be used in strong magnetic fields of order 10^{-4} T (1 gauss). They do not provide absolute measurements and need to be calibrated.

10.5.2 Alkali-vapour magnetometers

Alkali-vapour magnetometers depend on the Zeeman splitting of atomic energy levels in alkali vapour. They have proved more useful in space vehicles than proton precession magnetometers, which necessitate the switching on and off of a comparatively large magnetic field.

In principle any alkali vapour could be used but rubidium and caesium have proved the most convenient because they need be raised only to a comparatively low temperature to produce the required vapour pressure. Although alkali-vapour magnetometers do not measure the direction of the field they provide an absolute measure and do not need calibrating.

The principle of action can be illustrated with the help of fig. 10.10 which shows the relevant energy levels of an atom of Rubidium 85 in

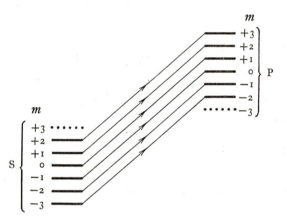

Fig. 10.10. To show, in schematic outline, how light with one sense of circular polarization is absorbed when it produces transitions from a Zeeman-split S-level to a Zeeman-split P level. Because Δm must be equal to $+1$, atoms in the S-level with $m = +3$ cannot absorb the light.

schematic outline. The nuclear spin quantum number $I = 5/2$ so that the ground (S) state is split into two levels with total spins 3 and 2 (hyperfine structure). In a magnetic field with induction B the level with total spin 3 becomes split by the Zeeman effect into close sub-levels having magnetic quantum numbers m from -3 to $+3$. The difference ΔE of energy between these sub-levels is related to the Larmor frequency† ($\Omega_L = eB/2m_e$) of an electron in the imposed magnetic field by the relation $\Delta E = h\Omega_L/2\pi$. The object of the method is to measure this frequency and hence to deduce B.

† The orbit of an electron rotating round a positive nucleus precesses round an applied magnetic field with the Larmor frequency.

Absorption of light can change the atom's energy from one of the S levels to one of the P levels which are split in a similar way and a change in the opposite direction is accompanied by emission of light. But a transition can occur only between an S and a P level if there is a change Δm of the magnetic quantum number by ± 1 or 0. If $\Delta m = \pm 1$ the light absorbed or emitted is circularly polarized with one sense or the other; if $\Delta m = 0$ it is linearly polarized.

In the magnetometer light from a lamp containing rubidium vapour is first circularly polarized and is then passed through a cell, also containing rubidium vapour, that is situated in the magnetic field to be measured. When the light is absorbed in the cell atoms are removed from the sub-levels of the ground state into the sub-levels of the P state, but, if the light is circularly polarized with the proper sense of rotation, the transitions occur only when $\Delta m = +1$, so that, as shown in fig. 10.10, no transition can take place out of the level $m = +3$ in the ground state, or into the level $m = -3$ in the P state.

The excited atoms in the P sub-levels now fall back to the S sub-levels, and radiation is emitted in the process, in particular some fall back to the S sub-level with $m = +3$ *from which they cannot be removed by the circularly polarized light*. Nor can atoms in this state spontaneously transfer to any other of the S sub-levels; they remain in a metastable state from which they can be removed only by some non-radiating process such as a collision or (as we shall see later) by being stimulated suitably from outside.

When the light is first turned on it is absorbed by producing transitions from the S to the P states. After some time, however, all the atoms that have made these transitions have fallen back to that S sub-level, with $m = +3$, from which the radiation can no longer remove them. The light is then no longer absorbed. If, somehow, the atoms could be removed from this metastable sub-level to one of the other sub-levels of the S state, with a different value of m, they could once again absorb the incident light. They can be stimulated to make that transition by acting upon them with a magnetic field that rotates at frequency Ω_L corresponding to the energy difference ΔE between the sub-states: but $\Omega_L = eB/2m_e$ so that if it is known the magnitude B of the imposed field can be determined.

In the magnetometer the field B, that produces the Zeeman

9

splitting, is the field that is to be measured; insertion of numerical values shows that the Larmor frequency, which is proportional to the field, is about $14\,\mathrm{Hz}$ for a field of $10^{-9}\,\mathrm{T}$ (1 gamma). The magnetic field required to stimulate the depopulation of the metastable state must thus rotate at this frequency. When it is applied the vapour is once more capable of absorbing the circularly polarized light. The measurement might therefore be made by applying a magnetic field of variable frequency and finding that frequency at which the absorption of the circularly polarized light was greatest. The applied field need not be circularly polarized, a linearly polarized field is sufficient because it is resolvable into two circularly polarized ones, one of which acts on the atoms to produce the required transitions.

It has proved convenient to make the measurement a little differently by causing the transparency of the vapour to vary at the Larmor frequency. This has been accomplished by arranging for the transmitted light to fall on a photocell to produce an e.m.f. that is fed back to control the strength of the applied field. A somewhat complicated arrangement then causes the whole system to oscillate at the Larmor frequency: this frequency is relayed to the ground to provide the desired measure of the imposed magnetic field.

10.6   Measurement of electric fields [87, 96]

If a satellite is moving through the steady magnetic field (B) of the earth with a velocity v it experiences an electric field $E = v \times B$: insertion of the magnitudes $v = 8\,\mathrm{km\,s^{-1}}$ and $B = 10^{-5}\,\mathrm{T}$ shows that $E = 8 \times 10^{-2}\,\mathrm{V\,m^{-1}}$. The existence of this induced field makes it difficult to measure the electrostatic fields, of the same order, that are believed to exist in different parts of the ionosphere. Before describing the methods that have been used to measure them it is desirable to enquire a little more fully into the precise meaning of the velocity v when the satellite is situated above the rotating earth.

Faraday showed that an electric field is induced at a point that moves in a circle round a bar magnet in a plane perpendicular to the axis of the magnet, as in fig. 10.11 (a), but that if the point is stationary and the magnet rotates about its axis, as in (b), there is no induced field. He also showed that if the magnet rotates while the point moves, the induced field depends only on the motion of the point and not at all on the rotation of the magnet. It follows that if there were no ionosphere the field $E = v \times B$ induced in a moving satellite would have to be calculated by using the absolute value of v, not the velocity relative to the surface of the earth. Thus, although a 'geostationary' satellite appears stationary to an earth-bound observer, it is moving with a velocity of about $3 \, \text{km s}^{-1}$ in a field whose induction is about $10^{-7} \, \text{T}$ and so, in the absence of an ionosphere, would experience an electric field of strength about $3 \times 10^{-4} \, \text{V m}^{-1}$: this field would be unaltered if the earth ceased to rotate.

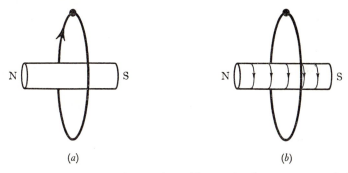

(a) (b)

Fig. 10.11. An electric field is experienced by a point that moves round a bar magnet, as at (a) but not by a stationary point near which a bar magnet rotates, as at (b).

The situation is, however, different when the conducting ionosphere is present. The electric field resulting from its movement through the earth's magnetic field produces a charge separation which sets up an electrostatic field just sufficient to cancel the field $v \times B$ at a point moving with the ionosphere, so that a geostationary satellite does *not* experience an electric field because of its motion. The only satellites that experience an induced field are those with velocity v relative to the ionosphere and they find $E = v \times B$. It follows that electric fields in the ionosphere (except the polarization field that results from its rotation with the earth) are most easily measured from a geostationary satellite.

It might be thought that electric fields in the vicinity of a satellite could be measured by finding the potential difference between separate probes extended from the satellite or by finding the charge flowing to and from a part of the surface that is alternately covered and uncovered. In experiments of that kind, however, it is very difficult to eliminate the effects of plasma sheaths and of possible changes in the work functions of the conductors. Some success has nevertheless been obtained, particularly with geostationary satellites.

The most valuable measurements of electric fields in the ionosphere have been obtained by observing the movements of luminous clouds of ionized gases ejected from rockets or satellites. The observations are made at the ground and it is usually arranged that the clouds of ionized gas are accompanied by luminous clouds of non-ionized gas ejected at the same time: the two can be distinguished because they emit light of different colours. The magnetic field of the earth causes the ionized and non-ionized gases to diffuse differently even when there is no electric field. If an electric field is present the ion cloud also drifts bodily, in a way that is discussed in §7.2.2. From this drift deductions are made about the electric field. Meanwhile bodily movements of the neutral atmosphere can be detected by observing the drift of the non-ionized cloud.

Appendix A

Chapman's theory of a production layer

A. i Results that apply to all atmospheres whatever their height-distributions

The following analysis is essentially the same as that of Chapman [66]:

Suppose the atmosphere to be horizontally stratified so that the concentration (n) of atoms (or molecules) depends only on the height (h) and that ionizing radiation of intensity I (energy flux per unit area) travels a distance s towards the earth in a direction making an angle χ with the vertical. Then if the absorption cross-section of an atom is σ the intensity (I) of the radiation decreases as given by

$$-dI/ds = \sigma n I \qquad (A\,1)$$

If C electrons are produced by each unit of energy the rate (q) of production in unit volume is

$$q = -C\,dI/ds = C\sigma n I \qquad (A.\,2)$$

it reaches a peak at a distance s (oblique) where

$$\frac{dq}{ds} = 0 = C\sigma\left(I\frac{dn}{ds} + n\frac{dI}{ds}\right)$$

or

$$\left(\frac{1}{n}\frac{dn}{ds}\right)_m + \left(\frac{1}{I}\frac{dI}{ds}\right)_m = 0 \qquad (A.\,3)$$

where the subscript m denotes the value at the peak of q. Now s (measured towards the earth) is related to h (measured upwards) by $s = -h\sec\chi$ so that

$$\frac{1}{n}\frac{dn}{ds} = \left(-\frac{1}{n}\frac{dn}{dh}\right)\cos\chi = \frac{\cos\chi}{H} \qquad (A.\,4)$$

where H is the distribution height of the gas. By making use of (A 1) and (A. 4), (A. 3) can be written

$$\sigma H_m n_m \sec\chi = 1 \qquad (A.\,5)$$

Now $H_m n_m$ is the total number of atoms in a superincumbent vertical column of unit cross-section; $H_m n_m \sec \chi$ is thus the total number $[N_m(\chi)]$ in an oblique column of unit cross-section, lying parallel to the direction of the incident radiation and having its base at the level of maximum production. (A. 5) can thus be written

$$\sigma N_m(\chi) = 1 \qquad (A.6)$$

If I_∞ denotes the intensity of the radiation before it reaches the atmosphere, (A. 1) shows that

$$\log(I/I_\infty) = -\sigma \int_s^\infty n\,ds = -\sigma N(\chi) \qquad (A.7)$$

But at the peak of production $\sigma N_m(\chi) = 1$ so that

$$I_m/I_\infty = e^{-1} \qquad (A.8)$$

The astronomers would say that the radiation has then penetrated to 'unit optical depth'.

The rate of production (q_m) at the peak is derived from (A. 2) by substituting for n_m from (A. 5) and for I_m from (A. 8) to give

$$q_m = (CI_\infty/eH_m)\cos\chi \qquad (A.9)$$

If electrons are lost by recombination at a rate $\alpha[e]^2$, where α is independent of height, and if the rate of loss balances the rate of production, then the peak concentration ($[e]_m$) of electrons is given by $q_m = \alpha[e]_m^2$ so that

$$[e]_m = (CI_\infty/\alpha eH)^{\frac{1}{2}}(\cos\chi)^{\frac{1}{2}} \qquad (A.9a)$$

A. ii   Results that apply to an atmosphere with an exponential height distribution

It is especially to be noticed that the previous calculations apply to a single absorbing gas whatever its height-distribution. Suppose next that the particles of the gas have a concentration represented by $n = n_0 \exp(-h/H)$, so that, from (A. 5), the peak of production occurs at the height (h_m), where

$$\sigma H n_0 \exp(-h_m/H)\sec\chi = 1$$

or

$$\exp(+h_m/H) = \sigma H n_0 \sec\chi \qquad (A.10)$$

The way in which the intensity I of the radiation depends on height (h) is now deduced as follows: From (A. 1)

$$-\,dI/ds = \sigma n I$$

or $\qquad dI/dh = \sigma n I \sec \chi = I \sigma n_0 \exp\left(-h/H\right)\sec \chi$

hence $\qquad \log\left(I/I_\infty\right) = -\,\sigma n_0 H \sec \chi \exp\left(-h/H\right)$

or $\qquad I = I_\infty \exp\left\{-\,\sigma n_0 H \sec \chi \exp\left(-h/H\right)\right\}$ $\qquad$ (A. 11)

The rate of production (q) is now obtained by substituting for I and n into (A. 2) to give

$$q = C\sigma n I$$
$$= C\sigma n_0 I_\infty \exp\left\{-h/H - n_0\,\sigma H \sec \chi \exp\left(-h/H\right)\right\}$$

It is convenient to transform this expression by substituting for CI_∞ from (A. 9) to give

$$q = q_m \,\sigma n_0 H \sec \chi \exp\left\{1 - h/H - n_0\,\sigma H \sec \chi \exp\left(-h/H\right)\right\}$$

and then writing $\sigma n_0 H \sec \chi = \exp\left(h_m/H\right)$ from (A. 10) to give

$$q = q_m \exp\left\{1 + \frac{h_m - h}{H} - \exp\left(\frac{h_m - h}{H}\right)\right\} \qquad \text{(A. 12)}$$

Now write $\qquad\qquad y = (h - h_m)/H \qquad\qquad$ (A. 13)

to represent heights measured from the level (h_m) of the peak, in terms of H as a unit, to obtain

$$q = q_m \exp\left\{1 - y - \exp\left(-y\right)\right\} \qquad \text{(A. 14)}$$

The simplicity of this expression shows that the production layers corresponding to all intensities (I_∞) of radiation, and all angles (χ) of incidence have the same form in terms of the normalized quantities y and q/q_m.

It is sometimes convenient to express q, not in terms of the quantities q_m and h_m at the peak of production when the direction of the incident radiation makes an angle χ with the vertical, but in terms of q_0 and h_0, the corresponding quantities when $\chi = 0$. Then, substituting

$$q_m = q_0 \cos \chi$$

(from A. 9) and $\exp(h_m/H) = \sec\chi\exp(h_0/H)$ (from A. 10) into (A. 12) gives

$$q = (q_0\cos\chi)\sec\chi\exp\left\{1 + \frac{h_0 - h}{H} - \sec\chi\exp\left(\frac{h_0 - h}{H}\right)\right\}$$

or, with
$$z = (h - h_0)/H \qquad\qquad\qquad (A.\,15)$$

as a new parameter to measure heights, in terms of H as a unit, from the level of the peak when $\chi = 0$

$$q = q_0\exp\{1 - z - \sec\chi\exp(-z)\} \qquad\qquad (A.\,16)$$

Appendix B

The time-constant for loss by diffusion

If a major atmospheric constituent is in equilibrium under gravity with scale height H_D and a minor constituent is distributed throughout it with concentration $n = n_0 \exp(-h/\delta)$, then (6.17) (p. 107) shows that

$$\frac{dn}{dt} = \gamma n$$

with

$$\gamma = D\left(\frac{1}{\delta} - \frac{1}{H_D}\right)\left(\frac{1}{\delta} - \frac{1}{H_N}\right)$$

where D is the diffusion coefficient and H_N is the scale height of the minor constituent (not necessarily equal to its distribution height).

Table 9 lists the approximate magnitudes of γ/D for several values of δ.

TABLE 9

Value of δ	Approximate magnitude of γ/D
$>H_D$ and $>H_N$	$1/H_D H_N$
$= H_N$	0
between H_D and H_N	$\begin{cases} -1/\delta\, H_D \text{ (if } H_N > H_D) \\ -1/\delta\, H_N \text{ (if } H_N < H_D) \end{cases}$
$= H_D$	0
$<H_D$ and $<H_N$	$1/\delta^2$
negative (upwards increase of n)	$>1/H_D H_N$

Under many circumstances δ, H_N and H_D are roughly of the same order so that then

$$\gamma \fallingdotseq D/(\delta^2 \text{ or } H_N^2 \text{ or } H_D^2) \tag{B.1}$$

This approximate result is useful when a comparison is to be made between rates of change of concentration resulting from (a) photochemical processes and (b) diffusion.

The table shows that there are two situations in which $dn/dt = 0$. One occurs when $\delta = H_N$ corresponding to the situation in which the

[239]

gas is distributed exponentially with its natural scale height. The velocity $W = -D\left(-\dfrac{1}{\delta} + \dfrac{1}{H_N}\right)$ (6.13) is then everywhere zero: the gas is at rest and the situation is one of equilibrium. The second situation occurs when $\delta = H_D$ and corresponds to one where the minor constituent is exponentially distributed with the distribution height of the major constituent: it is perhaps unexpected and deserves further consideration as follows.

When $\delta = H_D$ (6.13) shows that $W = -D\left(-\dfrac{1}{H_D} + \dfrac{1}{H_N}\right)$ and is not zero: it increases with height, like D. The number of minor constituent particles crossing unit area in unit time is, however, given by

$$nW = -nD\left(-\frac{1}{H_D} + \frac{1}{H_N}\right)$$

and the upwards *decrease* of $n[\propto \exp(-h/\delta)]$ is just sufficient to balance the upwards *increase* of $D[\propto \exp(h/H_D)]$ so that there is a steady flow of gas, upwards or downwards according as H_N is greater or smaller than H_D. In spite of the fact that $dn/dt = 0$ the situation is thus not one of equilibrium.

Appendix C

List of main symbols

B	magnetic induction
B_L	component of B along wave normal
B_T	component of B perpendicular to wave normal
C	number of ions produced by unit energy of XUV radiation
C_V	specific heat at constant volume
D	diffusion coefficient: electric displacement
d	geocentric distance to magnetopause
E	electric field
$\mathscr{E}$	e.m.f. $= \oint E \, \mathrm{d}l$ round a circuit
e	electronic charge
F	force
G	gravitational constant
g	acceleration of gravity
H	scale height: magnetic field
h	height: Planck's constant
I	intensity of ionizing radiation (power flux per unit area): dip angle
i	current
j	current density
K	heat conductivity: wave number (electro-acoustic waves)
k	Boltzman's constant: wave number (all waves except electro-acoustic): reaction rate
L	loss rate
M	magnetic dipole moment
m_e	mass of electron
m_i	mass of ion
N	number of particles in column of unit cross-section: particle concentration (only in chapter 8)
n	particle concentration: refractive index
P	volume polarization
p	pressure
Q	power (energy flux) of radiation per unit area

q	rate of production of electrons
R	radius of earth: resistance: sunspot number
r	radius of orbit
S	flux of solar decimetric radiation (units of $10^{-22}\,\mathrm{W\,m^{-2}\,Hz^{-1}}$)
$S(k)$	Fourier transform of $s(h)$
s	distance
$s(h).\mathrm{d}h$	incoherent power scattered per unit area from a thin layer of thickness $\mathrm{d}h$
T	temperature
t	time
U	heat energy
V	drift velocity
v	gas-kinetic velocity
W	vertical component of drift velocity: energy
X_e	$ne^2/\epsilon_0 m_e \omega^2$
X_i	$ne^2/\epsilon_0 m_i \omega^2$
Y_e	Ω_e/ω
Y_i	Ω_i/ω
Y_L	value of Y when $B = B_L$
Y_T	value of Y when $B = B_T$
y	$(h-h_m)/H$ (appendix A)
z	$(h-h_0)/H$ (appendix A)
[e, O, etc.]	concentration (no. per unit volume) of particles e, O, etc.
α	recombination coefficient: pitch angle: ratio T_e/T_i
β	linear loss coefficient: gradient of scale height
γ	rate coefficient for loss by diffusion
δ	distribution height
ϵ_0	permittivity of free space
ζ	displacement in electro-acoustical wave
θ	angle between wave normal and geomagnetic field
λ	wavelength: ratio $[O_2^-]/[e]$
μ	magnetic moment: real part of refractive index
μ_0	permeability of free space
ν	collision frequency: frequency of a quantum
ρ	mass density: geocentric distance
σ	atomic or molecular cross-section (various kinds): conductivity

ϕ	flux of induction
χ	solar zenith angle: attenuation constant
ψ	effective recombination coefficient in presence of negative ions: phase angle
ω	frequency (angular)
ω_{0e}, ω_{0i}	plasma frequency of electrons, ions $= (ne^2/\epsilon_0 m_e)^{\frac{1}{2}}, (ne^2/\epsilon_0 m_i)^{\frac{1}{2}}$
ω_0	$(\omega_{0e}^2 + \omega_{0i}^2)^{\frac{1}{2}}$
ω_{co}	cross-over frequency
Ω_e, Ω_i	gyro-frequency (angular) of electrons, ions, $= \|e\|B/m_e$, $\|e\|B/m_i$
Ω_L	Larmor frequency of precession

Conversion of units

In this book magnitudes are quoted in SI units; they can be converted to magnitudes and units in other systems as shown in the table below.

Quantity	Measure in SI units
Electron mass (m)	9.11×10^{-31} kg
Electron charge (e)	1.60×10^{-19} C
e/m	1.76×10^{11} C/kg
Electric permittivity of free space ϵ_0	8.85×10^{-12} F m^{-1}
Magnetic permittivity of free space μ_0	$4\pi \times 10^{-7}$ H m^{-1}
Boltzman's constant k	1.38×10^{-23} J degK^{-1}
Electron volt (eV)	1.6×10^{-19} J

Quantity	SI unit	Magnitude of SI unit in terms of other units
Wavelength	nm (nanometre)	10 Ångström
Density	kg m^{-3}	10^{-3} gm cm^{-3}
Force	N (newton)	10^5 dyne
Energy	J (joule)	10^7 erg
Energy density	J m^{-3}	10 erg cm^{-3}
Power flux	W m^{-2}	10^3 erg cm^{-2} s^{-1}
Spectral density of power flux	W m^{-2} nm^{-1}	10^4 erg cm^{-2} s^{-1} Å^{-1}
Recombination coefficient	m^3 s^{-1}	10^6 cm^3 s^{-1}
Diffusion coefficient	m^2 s^{-1}	10^4 cm^3 s^{-1}
Magnetic induction (B)	T (tesla)	10^4 gauss, 10^9 gamma

Bibliography

A. BOOKS AND CONFERENCE PROCEEDINGS

1 Akasofu, S-I. (1968). *Polar and magnetospheric substorms*, vol. 2. Astrophysics and Space Science Library. D. Reidel.
2 Alfven, H. (1950). *Cosmical electrodynamics*. Oxford University Press.
3 Budden, K. G. (1961). *Radio waves in the ionosphere*. Cambridge University Press.
4 Carovillano, R. L. *et al.* (1968), (editors). *Physics of the magnetosphere*. D. Reidel.
5 Chapman, S. and Bartels, J. (1940). *Geomagnetism*. Oxford University Press.
6 Chapman, S. (1964). *Solar plasma, geomagnetism and aurora*. Gordon and Breach. Also, *Geophysics, the earth's environment*, p. 373. Gordon and Breach (1962).
7 'CIRA'(1965). *Cospar International Reference Atmosphere*. North-Holland Pub., Amsterdam.
8 Clemmow, P. C. and Dougherty, P. J. (1969). *Electrodynamics of particles and plasmas*. Addison-Wesley.
9 Clemmow, P. C. (1966). *The plane wave spectrum representation of electromagnetic fields*. Pergamon.
10 Corliss, W. R. (1967). *Scientific Satellites*, NASA SP-133. U.S. Government Printing Office.
11 Craig, R. A. (1965). *The upper atmosphere, meteorology and physics*. Academic Press.
12 Danilov, A. D. (1970). *Chemistry of the ionosphere*. Plenum Press.
13 Davies, K. (1965). *Ionospheric radio propagation*. (Nat. Bur. Std. Monograph 80), U.S. Government Printing office.
14 Davies, K. (1969). *Ionospheric radio waves*. Blaisdell.
15 Denisse, J. F. and Delcroix, J. L. (1963). *Plasma waves*. No. 17 Interscience tracts on physics and astronomy. Wiley.
16 'Electromagnetic probing of the upper atmosphere' (1970). Special issue of *J. Atmos. Terr. Phys.* **32**, April.
17 Ginzburg, V. L. (1964). *The propagation of electromagnetic waves in plasmas*. Pergamon.
18 Helliwell, R. A. (1965). *Whistlers and related ionospheric phenomena*. Stanford University Press.
19 Hess, W. N. (1967). *The radiation belt and magnetosphere*. Blaisdell.
20 Hines, C. O., Paghis, I., Hartz, T. R. and Fejer, J. A. (1965), (editors). *Physics of the earth's upper atmosphere*. Prentice-Hall.
21 Jacobs, J. A. (1970). *Geomagnetic micropulsations*. Vol. 1 of Physics and Chemistry in Space. Springer-Verlag.
22 King, J. W. and Newman, W. G. (1967), (editors). *Solar terrestrial physics*. Academic Press.
23 McCormac, B. M. (1966), (editor). *Radiation trapped in the earth'*

magnetic field. Vol. 5 of Astrophysics and Space Science Library. D. Reidel.

24 Matsushita, S. and Campbell, W. H. (1967), (editors). *Physics of geomagnetic phenomena.* Academic Press.

25 Piddington, J. H. (1969). *Cosmic electrodynamics.* Wiley-Interscience.

26 Ratcliffe, J. A. (1959). *The magneto-ionic theory.* Cambridge University Press.

27 Ratcliffe, J. A. (1960), (editor). *Physics of the upper atmosphere.* Academic Press.

28 Rishbeth, H. and Garriott, O. (1969). *Introduction to ionospheric physics.* Academic Press.

29 Roederer, J. G. (1970). *Dynamics of geomagnetically trapped radiation.* Vol. 2 of Physics and Chemistry in Space. Springer-Verlag.

30 Stix, T. H. (1962). *The theory of plasma waves.* McGraw-Hill.

31 'Topside sounding of the ionosphere' (1969). Special issue of *Proc. IEEE* **57**, June.

32 White, R. S. (1970). *Space physics.* Gordon and Breach.

33 Whitten, R. C. and Poppoff, I. G. (1965). *Physics of the lower Ionosphere.* Prentice-Hall.

34 Whitten, R. C. and Poppoff, I. G. (1971). *Fundamentals of aeronomy.* Wiley.

B. INDIVIDUAL PAPERS

35 Allen, C. W. (1965). 'The interpretation of the XUV solar spectrum.' *Space Sci. Rev.* **4**, 91.

36 Allen, J. A. van (1968). 'Particle description of the magnetosphere.' *Physics of the magnetosphere*, p. 147, (ref. 4).

37 Appleton, E. V. and Piggott, W. R. (1952). 'The morphology of storms in the F2 layer of the ionosphere.' *J. Atmos. Terr. Phys.* **2**, 236.

38 Axford, W. I. (1967). 'The interaction between the solar wind and the magnetosphere.' *Aurora and airglow*, pp. 499–509, (editor McCormac). Reinhold.

39 Axford, W. I. (1968). 'Observations of the interplanetary plasma.' *Space Sci. Rev.* **8**, 331.

40 Axford, W. I. (1967). 'Magnetic storm effects associated with the tail of the magnetosphere.' *Space Sci. Rev.* **7**, 149.

41 Axford, W. I. (1970). 'A survey of interplanetary and terrestrial phenomena associated with solar flares.' *Intercorrelated satellite observations related to solar events*, p. 7, (editors Manno and Page). D. Reidel.

42 Axford, W. I. and Hines, C. O. (1961). 'A unifying theory of high-latitude geophysical phenomena and geomagnetic storms.' *Can. J. Phys.* **39**, 1433.

43 Bailey, D. K. (1959). 'Abnormal ionization in the lower ionosphere associated with cosmic-ray flux enhancements.' *Proc. Inst. Radio Engrs.* **47**, 255.

44 Banks, P. M. (1966). 'Collision frequencies and energy transfer.' *Planet Space Sci.* **14**, 1096.

45 Banks, P. M. (1969). 'The thermal structure of the ionosphere.' *Proc. IEEE* **57**, 258.
46 Barrington, R. E. (1969). 'Ionospheric ion composition deduced from (topside sounder) VLF observations.' *Proc. IEEE* **57**, 1036.
47 Barrington, R. E. and Thrane, L. (1962). 'The determination of D-region electron densities from observations of cross-modulations.' *J. atmos. terr. Phys.* **24.** 31.
48 Bates, D. R. (1970). 'Reactions in the ionosphere.' *Contemp. Phys.* **11**, 105.
49 Bauer, S. J. (1966). 'The constitution of the topside ionosphere.' *Electron density profiles in ionosphere and exosphere*, p. 270, (editor Frihagen). Wiley-Interscience.
50 Beard, D. B. (1967). 'The solar wind.' *Rep. Prog. Phys.* **33**, 409.
51 Belrose, J. S. (1965). 'The Lower Ionospheric Regions.' *Physics of the earth's upper atmosphere*, chapter 3, p. 46, (ref. 20).
52 Belrose, J. S. (1965). 'The ionospheric F region.' *Physics of the earth's upper atmosphere*, chapter 4, p. 73, (ref. 20).
53 Belrose, J. S. (1970). 'Radio wave probing of the ionosphere by the partial reflection of radio waves (from heights below 100 km).' *J. atmos. terr. Phys.* **32**, 567.
54 Bennett, W. H. (1950). 'Radio frequency mass-spectrometer.' *J. App. Phys.* **21**, 143, 723.
55 Block, L. P. (1967). 'Coupling between the outer magnetosphere and the high-latitude ionosphere.' *Space Sci. Rev.* **7**, 197.
56 Booker, H. G. and Smith, E. K. (1970). 'A comparative study of ionospheric measurement techniques.' *J. atmos. terr. Phys.* **32**, 467.
57 Boyd, R. L. F. (1968). 'Langmuir probes on spacecraft.' *Plasma diagnostics*, chapter 12, (editor W. Lochte-Holtgreven). North-Holland Pub. Amsterdam.
58 Brady, A. H. *et al.* (1964). 'Long lived effects in the D region after the high-altitude nuclear explosion of July 9, 1962.' *J. geophys. Res.* **69**, 1921.
59 Brice, N. M. (1967). 'Bulk motion of the magnetosphere.' *J. geophys. Res.* **72**, 5193.
60 Brice, N. M. and Smith, R. L. (1965). 'Lower hybrid resonance emissions.' *J. geophys. Res.* **70**, 71.
61 Brinton, H. C. *et al.* (1969). 'Implications for ionospheric chemistry and dynamics of a direct measurement of ion composition in the F2 region.' *J. geophys. Res.* **74**, 2941.
62 Carlson, H. C. (1966). 'Ionospheric heating by magnetic conjugate point photoelectrons.' *J. geophys. Res.* **71**, 195.
63 Carpenter, D. L. (1966). 'Whistler studies of the plasmapause in the magnetosphere. 1. Temporal variations in the position of the knee and some evidence on plasma motions near the knee.' *J. geophys. Res.* **71**, 693.
64 Carpenter, D. L. (1968). 'Recent research on the magnetospheric plasmapause.' *Radio Sci.* **3**, 719.
65 Carpenter, D. L. and Smith, R. L. (1964). 'Whistler measurements of electron density in the magnetosphere.' *Rev. Geophys.* **2**, 415.

66 Chapman, S. (1931). 'The absorption and dissociative or ionizing effect of monochromatic radiation in an atmosphere on a rotating earth.' *Proc. phys. Soc.* **43**, 26.

67 Chapman, S. (1956). 'The electrical conductivity of the ionosphere: A Review.' *Nuovo Cim.* **4** (10), Suppl. 4, 1385.

68 Chapman, S. and Ferraro, V. C. A. (1931–3). 'A new theory of magnetic storms. I. The initial phase.' *Terr. Magn. atmos. Elect.* **36**, 77, 171; **37**, 147, 421: II., 'The main phase.' *ibid.* **38**, 79.

69 Dessler, A. J. (1968). 'Solar wind interactions and the magnetosphere.' *Physics of the magnetosphere*, p. 65, (ref. 4).

70 Dessler, A. J. and Parker, E. N. (1959). 'Hydromagnetic theory of geomagnetic storms.' *J. geophys. Res.* **64**, 2239.

71 Donahue, T. M. (1966). 'The problem of atomic hydrogen.' *Ann. Geophys.* **22**, 175.

72 Dougherty, J. P. (1961). 'On the influence of horizontal motion of the neutral air on the diffusion equation of the F-region.' *J. atmos. terr. Phys.* **20**, 167.

73 Dougherty, J. P. and Farley, D. T. (1960). 'A theory of incoherent scattering of radio waves by a plasma.' *Proc. R. Soc. A* **259**, 79.

74 Dungey, J. W. (1963). 'The structure of the exosphere, or adventures in velocity space.' *Geophysics*, p. 505, (editors DeWitt *et al.*). Gordon and Breach.

75 Dungey, J. W. (1967). 'The theory of the quiet magnetosphere.' *Solar terrestrial physics*, p. 91, (ref. 22).

76 Dungey, J. W. (1968). 'Waves and particles in the magnetosphere.' *Physics of the magnetosphere*, p. 218, (ref. 4).

77 Eccles, D. and King, J. W. (1969). 'A review of topside sounder studies of the equatorial ionosphere.' *Proc. IEEE* **57**, 1012.

78 Eccles, D. and King, J. W. (1970). 'Ionospheric probing using vertical incidence sounding techniques.' *J. atmos. terr. Phys.* **32**, 517.

79 Evans, J. V. (1967). 'Ground-based measurements of atmospheric and ionospheric particle temperatures.' *Solar terrestrial physics*, p. 289, (ref. 22).

80 Evans, J. V. (1969). 'Theory and practice of ionospheric study by Thomson scatter radar.' *Proc. IEEE* **57**, 496.

81 Evans, J. V. and Gastman, I. J. (1970). 'Detection of conjugate photoelectrons at Millstone Hill.' *J. geophys. Res.* **75**, 807.

82 Farley, D. T. (1970). 'Incoherent scattering at radio frequencies.' *J. atmos. terr. Phys.* **32**, 693.

83 Fejer, J. A. (1960). 'Hydromagnetic wave propagation in the ionosphere.' *J. atmos. terr. Phys.* **18**, 135.

84 Fejer, J. A. (1965). 'Motions of ionization.' Ch. 7, p. 157 in *Physics of the earth's upper atmosphere* (ref. 20).

85 Fejer, J. A. (1970). 'Radiowave probing of the lower ionosphere by cross-modulation techniques.' *J. atmos. terr. Phys.* **32**, 597.

86 Ferguson, E. E. (1969). 'Laboratory measurement of F-region reaction rates.' *Ann. Geophys.* **25**, 815.

87 Föppl, H. *et al.* (1967). 'Artificial strontium and barium clouds in the upper atmosphere.' *Planet. Space Sci.* **15**, 357.

88 Frank, L. A. (1967).'On the extra-terrestrial ring current during magnetic storms.' *J. geophys. Res.* **72**, 3753.

89 Friedman, H. (1969). 'Energetic solar radiations', p. 36 in Vol. 4, *Annals of the IQSY* (Editor, Stickland). MIT Press.

90 Gardner, F. F. and Pawsey, J. L. (1953). 'Study of the ionospheric D-region using partial reflections.' *J. atmos. terr. Phys.* **3**, 321.

91 Garriott, O. K., Da Rosa, A. V. and Ross, W. J. (1970). 'Electron content obtained from Faraday rotation and phase path length variations.' *J. atmos. terr. Phys.* **32**, 705.

92 Gold, T. (1959). 'Motions in the magnetosphere of the earth.' *J. geophys. Res.* **64**, 1219.

93 Goldberg, R. A. (1969). 'A review of theories concerning the equatorial F2 region ionosphere.' *Proc. IEEE* **57**, 1119.

94 Gringauz, K. I. (1967). 'Rocket and satellite measurements of ionospheric and magnetic particle temperatures', p. 341 in *Solar terrestrial physics* (ref. 22).

95 Gurnett, D. A. *et al.* (1965). 'Ion cyclotron whistlers.' *J. geophys. Res.* **70**, 1665.

96 Haerendal, G. *et al.* (1967). 'Motion of artificial ion clouds in the upper atmosphere.' *Planet. Space Sci.* **15**, 1.

97 Hanson, W. B. (1962). 'Upper atmosphere helium ions.' *J. geophys. Res.* **67**, 183.

98 Hanson, W. B. and Patterson, T. N. L. (1963). 'Diurnal variation of the hydrogen concentration in the exosphere.' *Planet. Space Sci.* **11**, 1035.

99 Harris, I. and Priester, W. (1962). 'Time-dependent structure of the upper atmosphere.' *J. atmos. Sci.* **19**, 286.

100 Heikkila, W. J. and Axford, W. I. (1965). 'The outer ionospheric regions.' *Physics of the earth's upper atmosphere*, chapter 5, p. 96, (ref. 20).

101 Helliwell, R. A. (1968). 'Whistlers and VLF emissions.' *Physics of the magnetosphere*, p. 106, (ref. 4).

102 Herzberg, L. (1965). 'Solar optical radiation and its role in upper atmospheric processes.' *Physics of the earth's upper atmosphere*, p. 31, (ref. 20).

103 Hines, C. O. (1963). 'The relation between hydromagnetic waves and the magneto-ionic theory.' *Symposium on electromagnetic theory and antennas*, Copenhagen, p. 287. Pergamon.

104 Hines, C. O. (1964). 'Hydromagnetic motions in the magnetosphere.' *Space Sci. Rev.* **3**, 342.

105 Hines, C. O. and Reid, G. C. (1965). 'Theory of geomagnetic and auroral storms.' *Physics of the earth's upper atmosphere*, p. 334, (ref. 20).

106 Hinteregger, H. E. and Watanabe, K. (1962). 'Photoionization rates in the E and F regions.' *J. geophys. Res.* **67**, 3372.

107 Hinteregger, H. E. (1965). 'Absolute intensity measurements in the extreme ultraviolet spectrum of solar radiation.' *Space Sci. Rev.* **4**, 461.

108 Kaiser, T. R. and Tunnaley, J. K. E. (1968). 'Radiofrequency impedance probes.' *Space Sci. Rev.* **8**, 32.

109 King, J. W. and Fooks, J. L. (1968). 'Long-lasting storm effects in the ionospheric D-region.' *J. atmos. terr. Phys.* **30**, 639.

110 King-Hele, D. G. (1966). 'Methods of determining air density from satellite orbits.' *Ann. Geophys.* **22**, 40.

111 Kockarts, G. and Nicolet, M. (1932). 'Le problème aéronomique de l'helium et de l'hydrogene neutres.' *Annls. Géophys.* **18**, 269.

112 Kohl, H. and King, J. W. (1967). 'Atmospheric winds between 100 and 700 km and their effects on the ionosphere.' *J. atmos. terr. Phys.* **29**, 1045.

113 Lauter, E. A. and Knuth, R. (1967). 'Precipitation of high energy particles into the upper atmosphere at medium latitudes after magnetic storms.' *J. atmos. terr. Phys.* **29**, 411.

114 Levy, R. H., Petschek, H. E. and Siscoe, G. L. (1964). *Amer. Inst. Aeronaut. Astronaut. J.* **2**, 2065.

115 Lindzen, R. S. and Chapman, S. (1969). 'Atmospheric tides.' *Space Sci. Rev.* **10**, 3.

116 Mackenzie, E. C. and Sayers, J. (1966). 'A radio-frequency electron-density probe for rocket investigation of the ionosphere.' *Planet. Space Sci.* **14**, 731.

117 Maeda, H. (1968). 'Variation in geomagnetic field.' *Space Sci. Rev.* **8**, 555.

118 Mange, P. (1960). 'The distribution of minor ions in electro-static equilibrium in the high atmosphere.' *J. geophys. Res.* **65**, 3833.

119 Matsushita, S. (1967a). 'Solar quiet and lunar daily variation fields.' *Physics of geomagnetic phenomena*, p. 302, (ref. 24).

120 Matsushita, S. (1967b). 'Geomagnetic disturbances and storms.' *Physics of geomagnetic phenomena*, p. 793, (ref. 24).

121 Mitra, A. P. (1968). 'A review of D-region processes in non-polar latitudes.' *J. atmos. terr. Phys.* **30**, 1965.

122 Ness, N. F. (1967). 'Observations of the interaction of the solar wind with the geomagnetic field during quiet conditions.' *Solar terrestrial physics*, p. 57, (ref. 22).

123 Ness, N. F. (1970). 'Magnetometers for space research.' *Space Sci. Rev.* **11**, 459.

124 Nicolet, M. (1965). 'Nitrogen oxides in the chemosphere.' *J. geophys. Res.* **70**, 679.

125 Nicolet, M. (1960). 'The properties and constitution of the upper atmosphere.' *Physics of the upper atmosphere*, chapter 2, p. 17, (ref. 27).

126 Nicolet, M. (1961). 'Helium, an important constituent of the lower exosphere.' *J. geophys. Res.* **68**, 2263 (61).

127 Nicolet, M. (1961). 'Structure of the thermosphere.' *Planet. Space Sci.* **5**, 1.

128 Nicolet, M. and Aikin, A. C. (1960). 'The formation of the D region of the ionosphere.' *J. geophys. Res.* **65**, 1469.

129 Nicolet, M. and Mange, P. (1954). 'The dissociation of oxygen in the high atmosphere.' *J. geophys. Res.* **59**, 15.

130 Nishida, A. (1966). 'Formation of plasmapause by the combined action

of magnetospheric convection and plasma escape from the tail.'
J. geophys. Res. **71**, 5669.

131 Nishida, A. and Maezawa, J. (1971). 'Two basic modes of interaction between the solar wind and the magnetosphere.' *J. Geophys. Rev.* **76**, 2254.

132 Norton, R. B., Van Zandt, T. E. and Denison, J. S. (1963). 'A model of the atmosphere and ionosphere in the E and F1 regions.' *Proc. Intern. Conf. Ionosphere*, p. 26. Institute of Physics & Physical Society, London.

133 Obayashi, T. (1967). 'The interaction of the solar wind with the geomagnetic field during disturbed conditions.' *Solar terrestrial physics*, p. 106, (ref. 22).

134 Obayashi, T. and Nishida, A. (1968). 'Large-scale electrical field in the magnetosphere.' *Space Sci. Rev.* **8**, 3.

135 O'Brien, B. J. (1967). 'Inter-relations of energetic charged particles in the magnetosphere.' *Solar terrestrial physics*, p. 169, (ref. 22).

136 Olson, W. P. (1970). 'Variations in the earth's surface magnetic field from the magnetopause current system.' *Planet. Space Sci.* **18**, 1471.

137 Oya, H. and Obayashi, T. (1967). 'Rocket measurements of the ionospheric plasma by gyro-plasma probe up to 1800 km.' *Rep. Ionosph. Res. Japan* **21**, 1.

138 Oya, H. (1968). 'Theoretical prediction on discrimination of modified and hybrid plasma resonances.' *Rep. Ionosph. Res. Japan* **22**, 119.

139 Oya, H. (1970). 'Ionospheric plasma disturbances due to a moving space vehicle.' *Planet. Space Sci.* **18**, 793.

140 Paghis, I. (1965). 'Magnetic and ionospheric storms.' *Physics of the earth's upper atmosphere*, chapter 12, p. 271, (ref. 20).

141 Parker, E. N. (1967). 'The dynamical theory of gases and fields in interplanetary space.' *Solar terrestrial physics*, p. 45, (ref. 22).

142 Parker, E. N. (1968). 'Dynamical properties of the magnetosphere.' *Physics of the magnetosphere*, p. 3, (ref. 4).

143 Patterson, T. N. L. (1968). 'Escape of helium by non-thermal processes.' *Rev. Geophys.* **6**, 553.

144 Paul, W. *et al.* (1958). 'Das elektrische Massenfilter als Massenspektrometer und Isotropen trenner.' *Z. Phys.* **152**, 143.

145 Price, A. T. (1969). 'Daily variations of the geomagnetic field.' *Space Sci. Rev.* **9**, 151.

146 Priester, W., Roemer, M. and Volland, H. (1966). 'The physical behaviour of the upper atmosphere deduced from satellite drag data.' *Space Sci. Rev.* **6**, 707.

147 Ratcliffe, J. A. (1956). 'The formation of the ionospheric layers F1 and F2.' *J. atmos. terr. Phys.* **8**, 260.

148 Ratcliffe, J. A. and Weekes, K. (1960). 'The ionosphere.' *Physics of the upper atmosphere*, chapter 9, p. 378, (ref. 27).

149 Reid, G. C. (1965). 'Solar cosmic rays and the ionosphere.' *Physics of the earth's upper atmosphere*, chapter 11, p. 245, (ref. 20).

150 Rishbeth, H. (1967). 'A review of ionospheric F region theory.' *Proc. IEEE* **55**, 16.

151 Rishbeth, H. and Barron, D. W. (1960). 'Equilibrium electron distributions in the ionospheric F2 layer.' *J. atmos. terr. Phys.* **18**, 234.

152 Rycroft, M. J. and Thomas, J. O. (1970). 'The magnetospheric plasmapause and the electron density trough at the Alouette 1 orbit.' *Planet. Space Sci.* **18**, 65.

153 Salpeter, E. E. *et al.* (1965). 'Incoherent scatter from plasma oscillations in the ionosphere.' *Phys. Rev. Lett.* **14**, 579.

154 Sayers, J. (1970). 'In situ probes for ionospheric investigations.' *J. atmos. terr. Phys.* **32**, 663.

155 Shabansky, V. P. (1968). 'Magnetospheric processes and related geophysical phenomena.' *Space Sci. Rev.* **8**, 366.

156 Small, K. A. and Butler, S. T. (1961). 'The solar semidiurnal atmospheric oscillation.' *J. geophys. Res.* **66**, 3723.

157 Sonett, C. P. *et al.* (1968). 'The geomagnetic tail.' *Physics of the magnetosphere*, p. 461, (ref. 4).

158 Standards on wave propagation: Definition of terms (1950). *Proc. Inst. Radio Eng.* **38**, 1264.

159 Stoffregen, W. (1970). 'Electron density variations observed in the E-layer below an artificial barium cloud.' *J. atmos. terr. Phys.* **32**, 171.

160 Tarpley, J. D. (1970). 'The ionospheric wind dynamo I and II.' *Planet. Space Sci.* **18**, 1075, 1091.

161 Taylor, H. A. *et al.* (1969). 'Ion depletion in the high-latitude exosphere: simultaneous observations of the light ion trough and the VLF cut off.' *J. geophys. Res.* **74**, 3517.

162 Thomas, L. (1970). 'F2 region disturbances associated with major magnetic storms.' *Planet. Space Sci.* **18**, 917.

163 Thomas, L. (1971). 'The lower ionosphere.' *J. atmos. terr. Phys.* **33**, 157.

164 Timleck, P. L. and Nelms, G. L. (1969). 'Electron densities less than 100 electron cm^{-3} in the topside ionosphere. The "beat" method.' *Proc. IEEE* **57** (6), 1164.

165 Troitskaya, V. A. (1967). 'Micropulsations and the state of the magnetosphere.' *Solar terrestrial physics*, (ref. 22).

166 Troitskaya, V. A. and Gul'elmi, A. V. (1967). 'Geomagnetic micropulsations and diagnostics of the magnetosphere.' *Space Sci. Rev.* **7**, 689.

167 Waldmeier, M. (1961). 'The sunspot activity in the years 1710–1960.' Schulthess: Zurich.

168 Weil, H. and Yorks, R. G. (1970). 'OGO satellite wake structure deduced from antenna impedance measurements.' *Planet. Space Sci.* **18**, 901.

169 Wilcox, J. M. (1968). 'The interplanetary magnetic field. Solar origin and terrestrial effects.' *Space Sci. Rev.* **8**, 258.

170 Yonezawa, T. (1966). 'Theory of formation of the ionosphere.' *Space Sci. Rev.* **5**, 3.

LATE ADDITIONS

171 Evans, J. V. (1972). 'Ionospheric movements measured by incoherent scatter. A review.' *J. atmos. terr. Phys.* **34**, 175.

172 Ferguson, E. E. (1971). 'D-region ion chemistry.' *Rev. Geophys. and Space Phys.* **9**, 997.

173 Jacchia, L. G. (1971). 'Revised static models of the thermosphere and exosphere with empirical temperature profiles.' *Smithsonian Astrophysical Observatory Special Report* **332**.

174 Rishbeth, H. (1972). 'Thermospheric winds and the F-region. A review.' *J. atmos. terr. Phys.* **34**, 1.

175 Willis, D. M. (1971). 'Structure of the magnetosphere.' *Rev. Geophys. and Space Phys.* **9**, 953.

176 Willmore, A. P. (1970). 'Electron and ion temperatures in the ionosphere.' *Space Sci. Rev.* **11**, 607.

Index